SAMPSON TECHNICAL COLLEGE

**Automatic
Control
System
Technology**

Automatic Control System Technology

DANIEL P. SANTE
Erie Community College
Buffalo, New York

PRENTICE-HALL, INC., *Englewood Cliffs, New Jersey* 07632

Library of Congress Cataloging in Publication Data

SANTE, DANIEL P
   Automatic control system technology.

  Includes index.
  1. Automatic control.   I. Title.
TJ213.S116   629.8   79-21911
ISBN 0-13-054627-5

Editorial/production supervision and interior design: Nancy Moskowitz
Cover design: Edsal Enterprises
Manufacturing buyer: Gordon Osbourne

©1980 by Prentice-Hall, Inc., Englewood Cliffs, N.J.   07632

All rights reserved. No part of this book may be reproduced in any form or by any means without permission in writing from the publisher.

Printed in the United States of America

10 9 8 7 6 5 4 3 2 1

PRENTICE-HALL INTERNATIONAL, INC., *London*
PRENTICE-HALL OF AUSTRALIA PTY. LIMITED, *Sydney*
PRENTICE-HALL OF CANADA, LTD., *Toronto*
PRENTICE-HALL OF INDIA PRIVATE LIMITED, *New Delhi*
PRENTICE-HALL OF JAPAN, INC., *Tokyo*
PRENTICE-HALL OF SOUTHEAST ASIA PTE. LTD., *Singapore*
WHITEHALL BOOKS LIMITED, *Wellington, New Zealand*

*To the new generation of students
who will make the world an even better place in which to live.*

Contents

PREFACE   *xv*

1 BASIC CONTROL SYSTEM CONCEPTS   *1*

     1-1 Introduction   *1*
     1-2 Simple Systems and the Time Constant   *2*
     1-3 Simple Closed Loop System   *6*
     1-4 The Transfer Function   *8*
     Glossary   *11*
     Problems   *12*

2 AMPLIFIERS WITH FEEDBACK   *15*

     2-1 Introduction   *15*
     2-2 Positive Feedback and Instability   *23*

2-3 Phase Shift as a Function of Time Delay 25
Glossary 26
Problems 27

3 GRAPHICAL SOLUTION OF NETWORK RESPONSE 31

3-1 Introduction 31
3-2 The Decibel 32
3-3 The Time Constant and the Corner Frequency 33
3-4 The Octave and the Decade 36
3-5 The Universal Plot of a Single Lag Network 38

3-5-1 Normalizing 40

3-6 Solutions of Networks in Series 43

3-6-1 The Bode plot 43
3-6-2 The amplifier 44
3-6-3 Gain crossover frequency 47

Glossary 49
Problems 50

4 THE S OPERATOR AND THE LAPLACE TRANSFORM 53

4-1 Introduction 53
4-2 Basic Types of Networks 54

4-2-1 The lag network 54
4-2-2 The lead network 54
4-2-3 Miscellaneous networks 56

Contents ix

  4-3 The s Operator 58
  4-4 Network Solutions Using the s Operator 59
  4-5 The Laplace Transforms 62

    4-5-1 Theory of Laplace transforms 63

  4-6 Use of the s Operator and the Laplace Transforms 66
  4-7 The Series LCR Circuit 70
    Glossary 72
    Problems 73

5 SOLUTION OF COMPLEX TRANSFER FUNCTIONS 77

  5-1 Introduction 77
  5-2 Solution of Second Order Equations in Terms of s 78
  5-3 The Lead Network 80
  5-4 Special Transfer Functions 84
    Glossary 91
    Problems 91

6 CONTROL SYSTEM COMPONENTS 95

  6-1 Introduction 95
  6-2 Position Potentiometer 97
  6-3 The Position Transformer (Synchro) 99

    6-3-1 The linear synchro 101
    6-3-2 The rotary encoder 101

  6-4 Modulator/Demodulator 101

6-5 Control System Drive Motors   *104*

    6-5-1   *Servo motor (angular displacement type)*   104
    6-5-2   *The servo control valve (linear displacement type)*   105

6-6 The Transfer Function of a Rotating Type Servo Control Motor   *107*
6-7 Transfer Functions for Control System Components   *110*
6-8 The Flow Diagram of a Simple Control System   *112*
6-9 Phase Compensation   *113*
6-10 The Rate Generator   *114*
     Glossary   *116*
     Problems   *117*

7 OPEN LOOP SYSTEM ANALYSIS   *121*

7-1 Introduction   *121*
7-2 Position Control System Transfer Function   *123*

    7-2-1   *Bode plot analysis*   124

7-3 Gain Effects on System Performance   *130*

    7-3-1   *Dead space*   130
    7-3-2   *Effect of low system gain*   132

7-4 Effect of a Lead Network on the Phase Margin   *133*
7-5 Effect of Rate Feedback on the Phase Margin   *135*
7-6 Velocity Lag Error   *136*
     Glossary   *137*
     Problems   *138*

8 CLOSED LOOP SYSTEM ANALYSIS   141

8-1 Introduction   *141*
8-2 The Closed Loop Position Control System   *142*

   8-2-1 Adding a gear train   *143*

8-3 Steady State Analysis   *144*
8-4 Effect of Gain on Closed Loop Response   *147*

   8-4-1 Peak response M_p at ω_p *147*

8-5 The Damping Factor   *149*

   8-5-1 Natural resonance ω_n *150*
   8-5-2 Determining the peak frequency response ω_p *151*
   8-5-3 Determining the bandwidth ω_B of a closed loop system   *155*

Glossary   *159*
Problems   *160*

9 GRAPHICAL SOLUTION OF CLOSED LOOP SYSTEMS   163

9-1 Introduction   *163*
9-2 The Nichols Chart   *164*
9-3 Procedure for Using the Nichols Chart   *164*
9-4 Interpreting the Nichols Chart   *169*

   9-4-1 Determining K_0 from the Nichols chart   *171*
   9-4-2 Determining ω_p, M_p, ω_B *174*

Glossary   *177*
Problems   *177*

10 TRANSIENT ANALYSIS   *179*

    10-1   Introduction   *179*
    10-2   Effect of a Step Input on System Response   *180*

        10-2-1   Response time and overshoot   184
        10-2-2   Settling time   186
        10-2-3   Damped response to a unit step input   188

    10-3   Types of Resonant Systems   *192*
    10-4   Transient Analysis Using the Laplace Transforms   *193*

        10-4-1   The electrical resonant system   194
        10-4-2   The spring-mass system   195
        10-4-3   The rotating system   196

    10-5   The Basic Servo System   *197*
        Glossary   *198*
        Problems   *199*

11 DESIGN CONSIDERATIONS OF DYNAMIC SYSTEMS   *203*

    11-1   Introduction   *203*
    11-2   The Motor Transfer Function in Terms of its Speed/Torque Characteristics   *204*

        11-2-1   Motor and load considerations   205

    11-3   Basic Servo System with 100% Feedback   *209*
        Glossary   *215*
        Problems   *216*

APPENDIX A The Motor Transfer Function in Terms of Moment of Inertia and Viscous Damping   *219*

APPENDIX B Rate Feedback   *223*

APPENDIX C The Lead Network   *229*

APPENDIX D Phase Margin   *233*

APPENDIX E A Speed Control System   *237*

APPENDIX F The Laplace Transformations   *241*

Preface

Automatic Control Systems is one of the most exciting areas of study in electrical technology. It encompasses nearly all of the basic concepts learned in electrical and communication theory.

It is the complete logical development and analysis of dynamic systems. A slightly different approach is presented than what is normally used in control system analysis. It is based on the expansions of commonly known electronic and electrical phenomena such as resonance and damping, and relies heavily on a good understanding of the time constant. The electrical circuit elements are replaced later with their mechanical counterparts or equivalents, and the solution of electro-mechanical system performance is then completed.

This text does not follow the usual procedure for learning control system theory. I have found that the student of electrical technology can interpret the fundamental behavior of automatic control systems more easily when approached with electrical symbols and concepts, than when it is introduced with the usual spring-mass system with its newtons, meters and kilograms.

The step by step development of simple closed-loop systems is included in this text. It permits the student to see the integration of the subsystems (or blocks) into a complete closed-loop system and establishes the criteria for determining the stability or instability, frequency response, and gain characteristics of the overall system.

This course is geared to the college student in electrical technology who has a working knowledge of complex algebra and trigonometry—an introduction to calculus is most desirable but not absolutely necessary. The use of an electronic calculator is encouraged. Numerous transformations from rectangular to polar coordinates are required; both the natural logarithms for solutions of the exponential function e^{-x} and logarithms to the base 10 for conversion to decibels are called for. Of course, the trigonometric functions are needed to establish various phase angles which are very important in the study of any closed-loop system. A few approximations and generalizations are used to help clarify some of the more complex mathematical relationships. This is done with very little error to the final system solutions.

Laplace transformations, in their simplest form, (the use of the s operator to obtain solutions of complex networks which are in terms of $j\omega$) are used as a tool, in much the same way we use logarithms to simplify the solution of multiple stage amplifiers for the overall gain and frequency response characteristics. The interrelationship between electrical and mechanical oscillatory systems is made quite clear by the use of these Laplace transforms. The transition from electrical system behavior to mechanical system behavior becomes simply a matter of interchanging complimentary components. For example, the "inductance" is replaced by a "mass", the "capacitor" by a "spring constant" and the "resistor" by a "damping device".

This text provides for a basic working knowledge of Automatic Control Systems. At the same time, it bridges the gap that usually exists in relating the similarities between electrical and mechanical system phenomena.

In this process, the importance of resonance in its many forms is emphasized. On the whole, students obtain a final review of nearly all the basic concepts to which they have been exposed, integrated into an electro-mechanical model called an Automatic Control System.

DANIEL P. SANTE

Buffalo, New York

Automatic Control System Technology

1

Basic Control System Concepts

1-1 INTRODUCTION

Throughout history, there have been human beings who desired to do work much beyond the limits of their meager physical capabilities. Animals helped them move large masses, the lever gave them a mechanical advantage, the wheel provided mobility, and later the motor, with its rotating shaft, was capable of performing very large blocks of work in a relatively short time. As motors became more powerful and refined, better systems were needed to control them and so emerged the field of *automatic control systems*.

Today, the field of automatic control systems encompasses nearly every facet of our society. In manufacturing, it appears in automated production and process control operations; in the food industry, it is used to automatically blend and package food; the military applications are numerous and extend from the aiming of guns to control of radar systems to automatic guidance of missile systems and aircraft. Even our automobiles contain equipment to automatically maintain a preset speed as power requirements vary due to changing road conditions.

The applications of automatic control systems is almost unlimited. However, they all have one thing in common: the *output* response is automatically compared to an *input* command and the difference, called the *error*, then automatically readjusts the system in an effort to reduce the error to zero.

The goal of this text is to develop simple techniques for predicting the overall performance of a fully operating automatic control system. Block diagrams of each part of the system will be analyzed and sets of equations established for the behavior of these blocks. The equations, which will be called *transfer functions*, will then be combined and the overall system performance predicted by the use of various mathematical tools, including complex algebra, a graphical technique called the Bode plot, and the use of the *s* operator (a simple mathematical procedure related to the Laplace transforms), which is so useful in network solution.

Operational amplifiers, with the appropriate feedback, will be used to simulate each of the system blocks that make up a complete system. In this day of microprocessors, there is a tendency to consider the analog approach to system solutions to be obsolete. Actually, the analog model of a system makes it possible, by use of a simple "breadboard," to see and understand the dynamic behavior of a complete system or any of its subsystems; it provides instant access to any part of a system to observe directly the effects of changes in gain, corner frequencies, and phase compensation.

1-2 SIMPLE SYSTEMS AND THE TIME CONSTANT

A few simple motor systems with which we are all familiar are the vacuum cleaner, the bench grinder, and the dentist's drill. Such systems appear as shown in Figure 1-1. Closing S_1 supplies power to the motor in the form of a *step input* [Figure 1-1(b)] and it begins to rotate. It takes time for it to reach full speed because of the following physical attributes:

Sec. 1-2   Simple Systems and the Time Constant   3

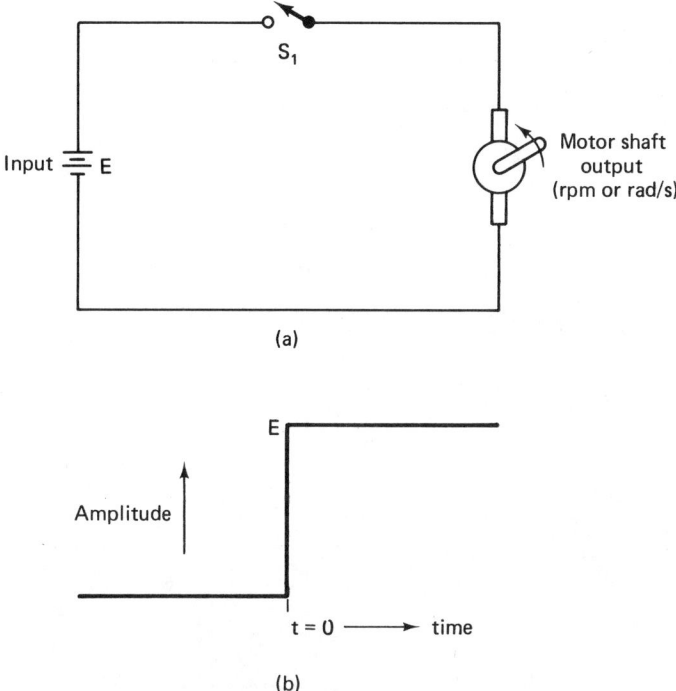

FIGURE 1-1 (a) Simple motor circuit; (b) step input of voltage.

1. Mechanical inertia resulting from the mass of the moving armature, which prevents the sudden change from standstill to full speed.
2. Electrical inertia is experienced in the field and armature coils as a result of the buildup of magnetic fields, which produce a counter electromotive force and a subsequent delay in the current transfer.
3. Friction and windage at the movable shaft slow down the rate of speed increase.

A plot of rpm versus time produces the response indicated in Figure 1-2. The response curve shown, which is a typical exponential function, is the same as the one obtained from the output voltage of a resistor–capacitor charging circuit after S_1 is closed. Note the similarity between Figures 1-2 and 1-3. Both responses are based on the time constant of the system involved.

4 Basic Control System Concepts   Ch. 1

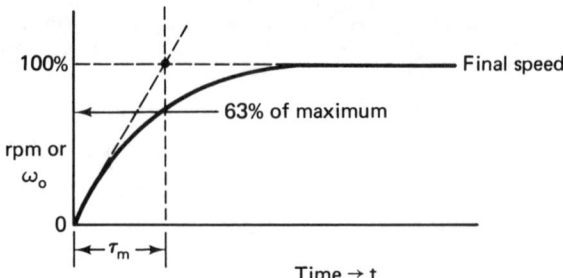

FIGURE 1-2 Buildup of motor speed with time.

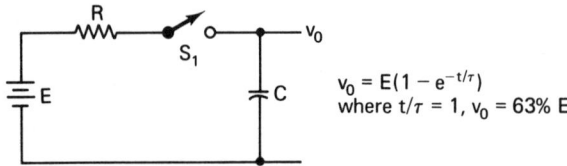

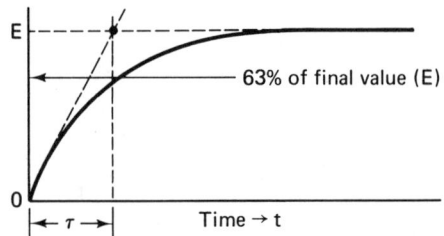

FIGURE 1-3 Buildup of voltage across C with time.

The *time constant* of a circuit is defined as the time required for the output, whether it be speed or voltage, to reach within $1/e$ of its final value, where e is the base of natural logarithms (i.e., 2.71828 ...). In the evaluation of the startup time of a motor, one finds this same exponential response. The time period to reach the same value (which is approximately 63% of the final speed) will be identified by the symbol τ_m and is referred to as the *motor time constant*. Both values rise in output following the same mathematical relationship, which describes the charging curve of the *RC* circuit:

$$v_0 = E(1 - e^{-t/\tau}) \qquad (1\text{-}1)$$

where $\tau =$ time constant RC
$v_o =$ instantaneous voltage at time t
$E =$ final output voltage

Similarly,

$$\omega_o = S(1 - e^{-t/\tau_m}) \qquad (1\text{-}2)$$

where $\tau_m =$ motor time constant
$\omega_o =$ instantaneous speed at time t
$S =$ final speed of motor

The rising exponential function,

$$y = (1 - e^{-t/\tau}) \qquad (1\text{-}3)$$

is normalized to unity by making t/τ equal the number of time constants, so that for $t/\tau = 1$,

$$y = (1 - e^{-1}) = 0.63 \quad \text{or} \quad 63\%$$

Similarly, the decaying exponential is described by

$$y = e^{-t/\tau} \qquad (1\text{-}4)$$

and for $t/\tau = 1$,

$$y = e^{-1} = 0.37 \quad \text{or} \quad 37\%$$

Both functions are summarized in Table 1-1 in terms of the number of time constants and are shown plotted in Figure 1-4, which is known as the *universal exponential curve* since the plot is applied both to mechanical and electrical systems that contain the time constant τ.

TABLE 1-1 Calculator value of e^{-x} and $(1 - e^{-x})$ for values of x from 1 to 4 where $x =$ number of time constants t/τ.

t/τ	$(1 - e^{-t/\tau})$	$e^{-t/\tau}$
1	63%	37%
2	$86\frac{1}{2}\%$	$13\frac{1}{2}\%$
3	95%	<5%
4	>98%	<2%

1-3 SIMPLE CLOSED-LOOP SYSTEM

Once the response characteristic of the motor is established, the motor can be controlled to perform many functions. For example, it can be used to position the tail surfaces of an aeroplane or missile, or for controlling the speed of a steel strip in a plating mill. With just an off–on switch, referred to as a "bang-bang" or "go/no-go" type of control, the degree of control is obviously meager and unpredictable. A technique for automatically comparing the output against the input (command signal) to assure that the output has responded as asked for by the input is much more desirable.

Thus, two systems are evident:

1. The input signal provides a command and the motor rotates to move the desired load, be it a tail surface or a strip of hot metal being milled, to a predetermined position without regard to evaluating the type of response obtained. This is called the *open-loop system* and is depicted in Figure 1-5.

2. A more sophisticated system is one in which an input/output comparison is made to determine if the output is responding to the input command. In the second system, the difference between the actual output displacement and the input command is termed *error*, and this difference signal provides the stimulus for motor operation until it is reduced to zero. This type of feedback system is called a *closed-loop system* and is depicted in Figure 1-6.

In Figure 1-5, a command signal causes the motor to rotate and the tail to move "up" or "down." To stop the tail from moving up or down requires that the input signal be removed. This method of control is too limited and is seldom used. A more practical approach is the closed-loop system of Figure 1-6. When an input signal (command) is given to move the tail surface, the amount of tail movement that results produces a voltage at the pick-off potentiometer (y) proportional to displacement, which is then used as feedback and automatically compared with the command signal to determine whether the system has performed its function of controlling the motor. When

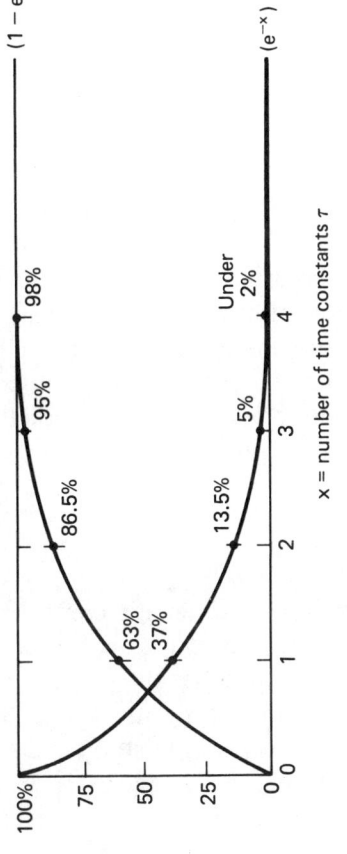

FIGURE 1-4 Universal exponential curves.

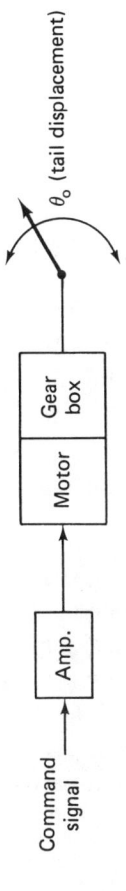

FIGURE 1-5 Open-loop system.

8 Basic Control System Concepts Ch. 1

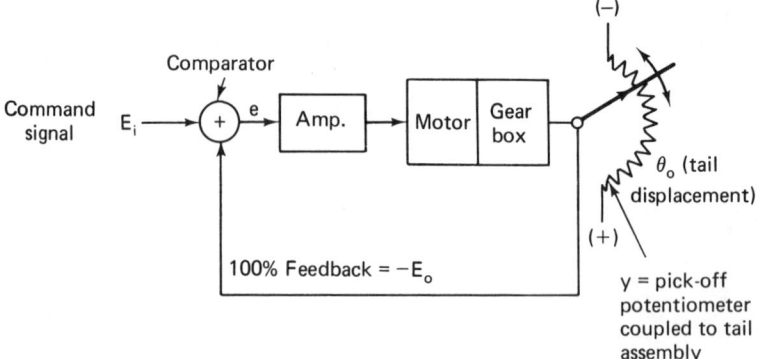

FIGURE 1-6 Closed-loop system.

the output voltage E_o is equal and opposite to the command signal E_i, the error voltage reduces to near zero and the motor stops because there is no longer drive power from the motor amplifier. Since the output is always directly proportional to the input signal in this case, the amount of displacement is easily predictable. It is customary to refer to an automatic closed-loop control system of this type (where a given command results in a predictable position output) as a *servo system*.

1-4 THE TRANSFER FUNCTION

The various elements that make up an automatic control system must be sensitive to the minute variations in signal levels (both command and feedback), and therefore amplifiers are used to improve system performance. Since the total gain used has a great effect on system performance, it behooves us to study in detail the amplifier with and without feedback. An amplifier is an active network wherein a change in the input, whether it be a voltage, current, or power, results in a change in the output.

The amplification characteristics of a device is usually identified in terms of its output/input ratio and is depicted mathematically as

$$\frac{\Delta E_o}{\Delta E_i} \quad \text{or} \quad \frac{\Delta I_o}{\Delta I_i} \quad \text{or} \quad \frac{\Delta P_o}{\Delta P_i} \qquad (1\text{-}5)$$

Amplifiers appear in many forms. They can be built with semiconductors, tubes, magnetic amplifiers, or combinations, to provide any gain desired.

In some special cases, such as the field-effect transistor, the output/input relationship involves different units, so the output/input relationship is not always a constant. Instead, we have a transfer characteristic that is not a simple ratio but is milliampere output/voltage input called (transconductance). The output/input relationship in different units is an important concept in control system analysis and is referred to as the *transfer characteristic* of the device.

In a typical servo system, electronic amplifiers are used to drive a motor, hydraulic valve, or magnetic clutch. The output may be in the form of degrees, inches, or revolutions per second, while the input can be the voltage level produced by a transducer, sensor, or any of the many forms in which a command signal can be given.

EXAMPLE 1-1

Given Figure 1-7, with a change in input of 1 V ($\Delta E_i = 1$ V), the linear displacement of the output in inches ($\Delta D =$ inches) provides the output/input relationship of 2 in./V for the system shown, *not* simply an output/input ratio of 2.

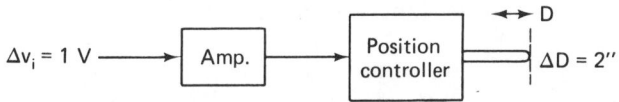

FIGURE 1-7 Linear displacement vs. voltage.

In the same way:

EXAMPLE 1-2

In Figure 1-8, the output/input relationship is 4°/V. Not to confuse system gain with the usual output/input relationship of V/V or mA/mA or W/W, which has no units, it becomes convenient to refer to the input/output relationship as the *transfer function*,*

---

* To complete the definition of a transfer function, it must include the effect of frequency on the amplitude characteristics. This frequency portion of the transfer function will be added after completion of the section dealing with networks.

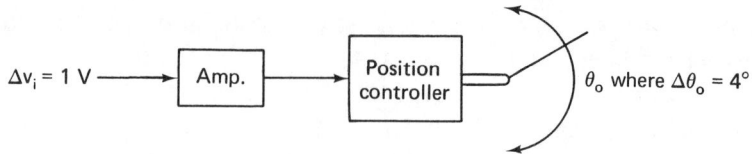

FIGURE 1-8 Rotary or angular displacement vs. voltage.

as, for example, TF = 2 in./V or 4°/V. A complete system could be a series of blocks as shown in Figure 1-9. Note that in each block, the letter K is used to signify the transfer function of each block that makes up the system and a subscript is used to indicate whether it is an amplifier (K_a), a motor (K_m), or a controller (K_c).

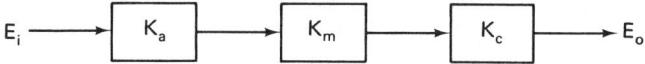

FIGURE 1-9 Simple system where TF = E_o/E_i.

In the figure,

K_a is a voltage amplifier = E_a/E_i = V/V

K_m is a motor whose sensitivity = rpm/E_a = rpm/V

K_c might be a controller whose characteristic = E_o/rpm

The total transfer function becomes

$$\text{TF} = K_a \times K_m \times K_c \qquad (1\text{-}6)$$

Substituting for K_a, K_m, and K_c:

$$\text{TF} = \frac{E_a}{E_i} \cdot \frac{\text{rpm}}{E_a} \cdot \frac{E_o}{\text{rpm}} \qquad (1\text{-}7)$$

After performing multiplication, the rpm and E_a factors cancel out, leaving the transfer function of the system as

$$\text{TF} = \frac{E_o}{E_i} = \text{volts output per volt input} \qquad (1\text{-}8)$$

As equation (1-8) indicates, Figure 1-9 could be replaced with a simple amplifier with a voltage gain of E_o/E_i (recall that to complete the transfer function, the frequency characteristic must be included). This second part of the transfer function will be added after we have developed a good understanding of network behavior as frequency or time delay is varied.

GLOSSARY

Automatic Control System: A closed-loop system in which the output response is compared to the input command and corrected until the difference is near zero, so that for a given command signal, a predictable output is obtained.

Closed-Loop System: A system in which all or part of the output is fed back to the input. It is also referred to as a system with feedback.

Command Signal: The input signal to an automatic control system. In this text, it will be either an input angle θ_i or an input voltage E_i or v_i.

Comparator: More often referred to as the "summing junction." It is where the output voltage (which is being fed back to the input) is added to the input. The output of the comparator provides the error voltage (e), which drives the output amplifier.

Error Signal: The output from the comparator or summing junction.

Exponential Function: An output response expressed as an exponent in terms of a designated power of e (the base of natural logarithms), such as e^{-x} for a decaying characteristic and $(1 - e^{-x})$ for a rising characteristic.

Feedback: The technique or process of adding all or part of the output signal back to the input of a system. It is negative feedback when it cancels out part of the input signal and positive when it increases or adds to the input signal.

Open-Loop System: Any system in which an input signal results in an output response and the output in no way appears or affects the input. An example is the simple system using a switch to apply or remove power to a motor (i.e., an electric drill).

Servo System: An automatic control system in which the output (usually angular position) depends on the level of the command signal. The transfer function is either θ_o/θ_i or θ_o/E_i.

Summing Junction: Often referred to as SJ (*see* Comparator).

Tachometer: A voltage generator whose output voltage is directly proportional to rotational speed. The transfer function would be simply E_o/rpm.

Transfer Function: The output/input relationship of a component or system expressed as a ratio. (In later sections, it will also include the effect of frequency on the output amplitude.)

PROBLEMS

1.1 Given $y = e^{-x}$, solve for y when $x = 0.1, 0.5, 1, 2,$ and 3.

1.2 After finding y for $x = 1$, solve for y when $x = 3$, using the "repeating-time-constant approach" (i.e., $e^{-1} \cdot e^{-1} \cdot e^{-1}$).

1.3 Given $y = 1 - e^{-t/\tau}$, solve for y when $t = 0.1, 0.5, 1, 2,$ and 3 with $\tau = 1$ s.

1.4 Make rough plots for Problems 1.1 and 1.3.

1.5 If a motor takes 0.8 s to reach 63% of its final rated speed after power is applied, what is its time constant?

1.6 Given:

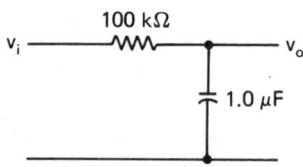

(a) What is the time constant of this circuit?
(b) What is the value of v_o after one time constant?
(c) Determine v_o at the end of 0.2 s.

1.7 A tachometer produces 4 V output when driven at a speed of 500 rpm. What is the transfer function of the tachometer?

1.8 If a rudder moves $+20°$ with the application of a $+5$ V command signal, what is the overall transfer function of the system?

1.9 If the application of 10° clockwise movement of a steering wheel results in a 30° displacement of the wheels, what is the gain of the system?

1.10 If the output speed of a motor is 2000 rpm with 10 V applied:
   (a) What is the transfer function of the motor?
   (b) Express your answer in rad/s/v.

1.11 What is the time constant of a system if it requires 0.2 s for the output (whether it is motor speed or voltage) to reduce to one-half of its initial condition?

1.12 How long will it take the output of an open-loop system to reach three-fourth of its final value if it has a built-in time constant of 14 s?

1.13 An important equation describing a system response is given by

$$A = e^{-\zeta \pi / \sqrt{1-\zeta^2}}$$

   (a) Determine A when $\zeta = 0.2$.
   (b) Determine A when $\zeta = 0.4$.

2

Amplifiers With Feedback

2-1 INTRODUCTION

Automatic control systems require some form of *feedback* (as mentioned in Chapter 1), where all or part of the output is fed back and compared (or added) to the input. Before developing the subject of feedback, it should be pointed out that most output-sensing devices used in control work (usually referred to as "end instruments") are energy transducers, such as accelerometers, position indicators, tachometers, and the like, which are voltage-output devices. Therefore, the voltage amplifier is an important part of control system study. At the same time, the voltage amplifier lends itself readily to the development of the fundamental equations for solving the output/input relationship of a closed-loop system. The analog computer (a valuable instrument for studying automatic control system behavior) is made up basically of many voltage amplifiers in cascade (series), each using some form of feedback to modify the amplitude and phase characteristics of each of the blocks that make up the complete system.

When an amplifier is specifically built for use in feedback systems, it will have a high voltage gain with a wide frequency response and will provide an output that is either in phase or is 180° out of phase with the input signal, depending on which set of input terminals are used. Provisions are included for feeding all or part of the output back to the input. These amplifiers are usually referred to as *operational amplifiers*, abbreviated op-amps. Numerous types of op-amps are available commercially, including prepackaged integrated circuits (ICs).

Symbolically, the operational amplifier is shown in Figure 2-1 (not including the power supply inputs) as a block with two inputs, v_i and v_f, and a single output, v_o. A positive signal into v_i (the +

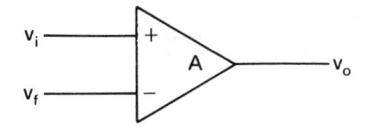

FIGURE 2-1 Symbol for an operational amplifier.

terminal) results in an output at v_o being in phase with the input; conversely, a phase shift of 180° results between an input into v_f (−) and the output v_o. By feeding a percentage of the output (which we will call β) back into the inverting input v_f, a valuable relationship develops (see Figure 2-2).

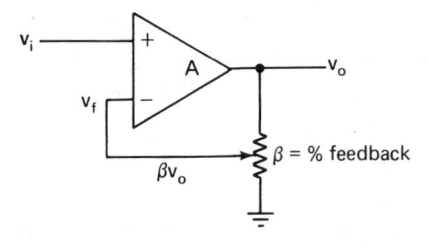

FIGURE 2-2 Op-amp with voltage feedback.

1. Without feedback (Figure 2-1)

$$v_o = A(v_i - v_f) \qquad (2\text{-}1)$$

2. With feedback (Figure 2-2)

$$v_f = \beta \cdot v_o \qquad (2\text{-}2)$$

Sec. 2-1 Introduction 17

3. Eliminating v_f and solving for v_o, we obtain

$$v_o = A(v_i - \beta v_o) = Av_i - A\beta v_o \tag{2-3}$$
$$v_o(1 + A\beta) = Av_i \tag{2-4}$$
$$\frac{v_o}{v_i} = \frac{A}{1 + A\beta} = G \tag{2-5}$$

In the foregoing analysis, when all or part of the output is fed back directly to the input, so that the output cancels part of the input [as shown by equation (2-3)], the amplifier is said to have *negative feedback*. Recognizing that v_o/v_i is the gain of the total system when feedback is used, we designate this as (G) to differentiate between the open-loop gain (A) of an amplifier without feedback.

When A is very large (10^5 or higher in typical IC packages such as the 741 series), a very accurate approximation of the closed-loop gain (G) is obtained by dividing the numerator and the denominator of equation (2-5) by A:

$$G = \frac{A/A}{(1/A) + (A\beta/A)} \quad \text{or} \quad G = \frac{1}{1/A + \beta}$$

With $A = 100{,}000$, $1/A = 10^{-5}$ and

$$G \approx \frac{1}{\beta} \tag{2-6}$$

A very valuable type of amplifier is obtained when $\beta = 1$. The noninverting input provides an output equal to the input but at a very low impedance (several hundred ohms or less), as contrasted to its very high input impedance (many megohms). It is called an *isolation amplifier* and is used to minimize loading of a very high impedance driving source by the lower input impedance of an op-amp, which is equal to the input resistor R_i.

EXAMPLE 2-1

Given an amplifier with an open-loop gain of 100, 1000, and 10,000 and a feedback ratio of $\beta = 10\%$, determine the exact closed-loop gain and compare this to the approximate gain of simply

$$G \approx \frac{1}{\beta} = \frac{1}{0.1} = 10$$

SOLUTION:

For $A = 100$,

$$G = \frac{A}{1+A} = \frac{100}{1+100(0.1)} = \frac{100}{11} = 9.09$$

for $A = 1000$,

$$G = \frac{1000}{1+1000(0.1)} = \frac{1000}{101} = 9.9$$

and for $A = 10,000$,

$$G = \frac{10,000}{1+10,000(0.1)} = \frac{10,000}{1001} = 9.99 \approx 10$$

Note that as A increases, G more nearly approaches $1/\beta$. Since it is not uncommon to use operational amplifiers with gains greater than 10^5 in actual control system designs, it is acceptable to use $G = 1/\beta$ as the closed-loop gain.

A more conventional approach in automatic control systems is to use a summing amplifier, commonly referred to as the *summing junction*, to combine the output with the input. The summing amplifier is adjusted for a gain of unity to provide a single output (e) equal to the sum of the inputs.

Again, the operational amplifier handily provides this function. By using current feedback as shown in Figure 2-3, the following is evident:

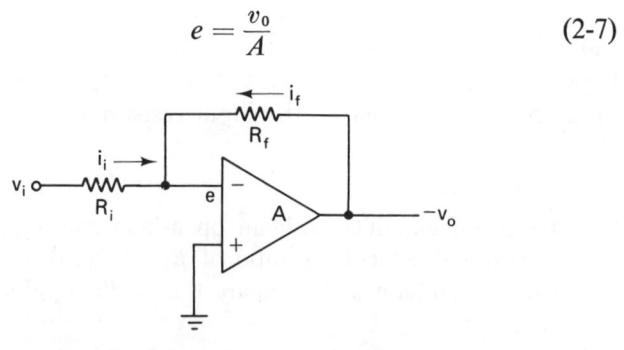

$$e = \frac{v_o}{A} \qquad (2\text{-}7)$$

FIGURE 2-3 Op-amp with current feedback.

Sec. 2-1　　　　　　　　　　　　　　　　Introduction　　**19**

where v_o is usually limited by the power supply to ± 10 V peak to peak or to 3.5 V rms. With $A > 10{,}000$,

$$e = \frac{3.5}{10^4} \text{ or less than } 3.5 \times 10^{-4} \text{ V/rms}$$

Thus, i_i (assuming that $e \approx 0$)

$$i_i = \frac{v_i - e}{R_i} \quad \text{or} \quad \frac{v_i}{R_i} \tag{2-8}$$

and i_f must equal

$$i_f = \frac{-v_o + e}{R_f} \quad \text{or} \quad \frac{-v_o}{R_f} \tag{2-9}$$

Since the voltage at e approaches zero and the input to the op-amp is a high impedance, the current at the junction must also approach zero, so that

$$i_i - i_f = 0 \quad \text{or} \quad i_f = i_i \tag{2-10}$$

from which

$$\frac{v_o}{v_i} = -\frac{R_f}{R_i} \tag{2-11}$$

Incidentally, the input impedance for the amplifier of Figure 2-3 is equal to R_i, whereas that of Figure 2-2 is many megohms, as mentioned earlier. The procedure just described can be extended to include several inputs, as shown in Figure 2-4.

By making $R_1 = R_2 = R_3 = R_f$,

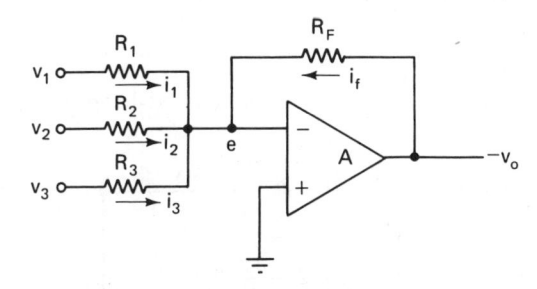

$$i_1 = \frac{v_1}{R_1}, \quad i_2 = \frac{v_2}{R_2}, \quad i_3 = \frac{v_3}{R_3}, \quad \text{and} \quad i_f = \frac{-v_o}{R_f} \tag{2-12}$$

FIGURE 2-4 Op-amp used as a summing amplifier.

20 Amplifiers With Feedback Ch. 2

Now $i_1 + i_2 + i_3 - i_f = 0$ [see equation (2-10)] and

$$\frac{v_1}{R_1} + \frac{v_2}{R_2} + \frac{v_3}{R_3} = \frac{v_o}{R_f} \quad \text{or} \quad \frac{1}{R}(v_1 + v_2 + v_3) = \frac{-v_o}{R} \quad (2\text{-}13)$$

so that

$$-v_o = v_1 + v_2 + v_3 \quad (2\text{-}14)$$

The summing amplifier is used to add various forms of feedback to the input signal to provide the error signal. Generally, the summing amplifier of Figure 2-4 is replaced with the symbols shown in Figure 2-5, where addition and subtraction take the form indicated in the figure.

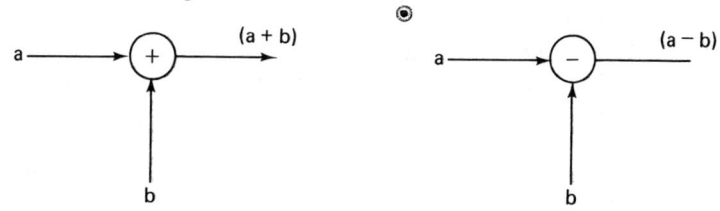

FIGURE 2-5 Summing junctions (symbols for addition and subtraction).

The overall system, showing direction and polarity of signals, is called the *block diagram*. Figure 2-6 is an example of a block diagram showing the signal paths of a system with feedback. Note in Figure 2-6 that the summing amplifier is (+) and A is (−), which indicates the feedback is negative. Then from Figure 2-6, we obtain:

$$y = A(x - \beta y) = Ax - A\beta y$$

and

$$y + A\beta y - Ax = y(1 + A\beta)$$

from which

$$\frac{y}{x} = \frac{A}{1 + A\beta} = G \quad \text{(the closed-loop gain)} \quad (2\text{-}15)$$

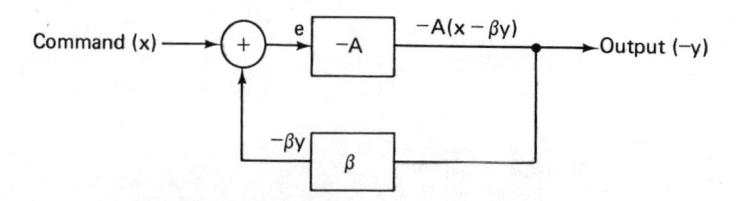

FIGURE 2-6 Block diagram of a closed-loop system.

If it is desired to use an amplifier (*A*) with no phase shift, the needed 180° phase shift of the output for negative feedback is accomplished in the feedback loop by making *β* minus quantity to show that the feedback is negative. The overall general feedback loops for both negative and positive feedback are shown in Figure 2-7.

This technique for describing the closed-loop response can be extended to include the addition of external disturbances, as, for example, the effect on an aircraft autopilot system caused by sud-

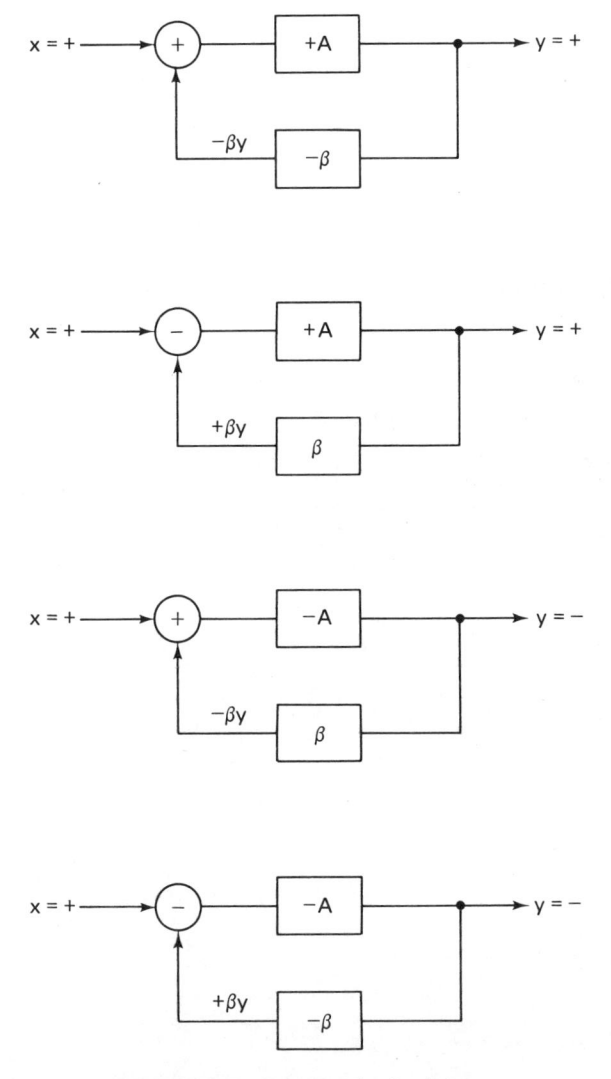

FIGURE 2-7 Negative feedback loops.

den wind gusts or other interference items, which will be called d. The block diagram including the disturbance d with negative feedback becomes that shown in Figure 2.8. Using the procedure developed,

$$y = A(x - \beta y + d) \qquad (2\text{-}16)$$

Example 2-2 will use equation (2-16) to show the effect of external disturbance d on a simple system similar to that of Figure 2-8.

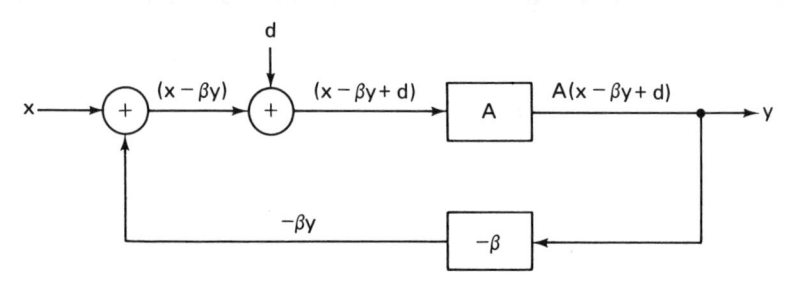

FIGURE 2-8 Closed-loop system with disturbance (d).

EXAMPLE 2-2

Let us make the disturbance d the same level as the command x, but first, solve for the output y with $d = 0$, then, for comparison, solve for y with $d = 2$:

$$x = 2 \text{ V}$$
$$d = 2 \text{ V}$$
$$A = 1000$$
$$\beta = 20\% \quad \text{or} \quad 0.2$$

SOLUTION:

First, with $d = 0$,

$$y = x \cdot \frac{A}{1 + A\beta} = 2 \cdot \frac{1000}{1 + 0.3(1000)} = 9.95 \text{ V}$$

Note this is very close to

$$G \approx \frac{1}{\beta} = \frac{1}{0.2} = 5 \quad \text{and} \quad y = 5 \times 2 = 10 \text{ V}$$

Sec. 2-2 Positive Feedback and Instability

Now with the disturbance $d = 2$, y will change to

$$y = A(x - \beta y + d)$$
$$= 1000(2 - 0.2y + 2)$$
$$= 1000(4 - 0.2y) = 4000 - 200y \qquad (2\text{-}17)$$
$$= \frac{4000}{201} = 19.9 \text{ V}$$

In Example 2-2 the disturbance d radically affects the output y. On the other hand, it can be shown that the disturbance d can be greatly reduced by rearranging the system blocks as shown in Figure 2-9. The output response y becomes

$$y = A(x - \beta y) + d \qquad (2\text{-}18)$$

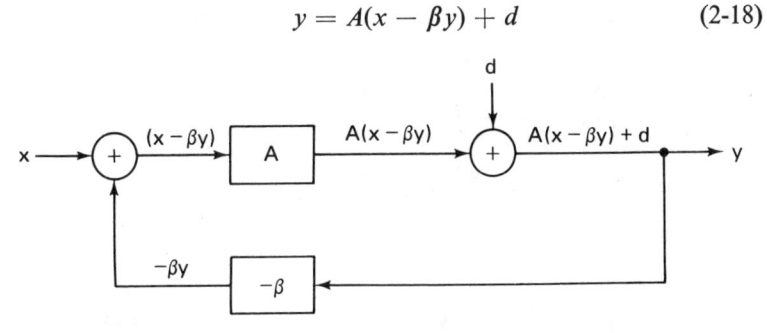

FIGURE 2-9 Closed-loop system with disturbance (d) inserted down the line.

Substituting values mentioned with $d = 2$,

$$y = 1000(2 - 0.2y) + 2$$
$$= 2000 - 200y + 2$$
$$= \frac{2002}{201} = 9.96 \text{ V}$$

In the second case, the disturbance has little effect on the output, which was 9.95 V without the disturbance and increased to 9.96 V with a disturbance of $d = 2$.

2-2 POSITIVE FEEDBACK AND INSTABILITY

In Section 2-1, the feedback was assumed to be negative with $G = A/(1 + A\beta)$. As the percent feedback increased, the closed-loop gain decreased. If an amplifier is built so that the portion of the output that is fed back to the input is in phase with the command signal, simple logic will indicate that the amplifier input will increase as the output increment is added in phase to the input. The output will continue to increase until the amplified output reaches saturation (Δv_o is reduced to zero), and the gain $\Delta v_o/\Delta v_i$ then becomes zero. At this point, the amplifier begins to recover until gain is restored. Immediately, the input begins to build up again until saturation is again encountered. The frequency at which this occurs depends on the time constants of the RC networks that make up the amplifier circuits. This phenomenon of repeated changes in output level is termed *oscillating* and defines a condition of instability. The multivibrator circuit of Figure 2-10 is a good example of this phenomenon put to good use.

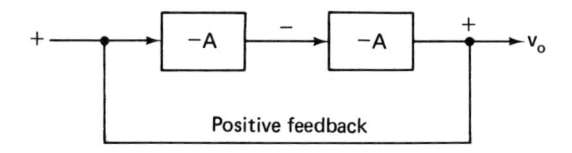

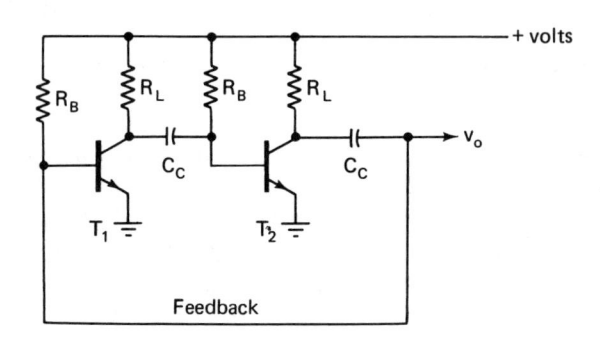

FIGURE 2-10 Example of positive feedback.

When does negative feedback become positive feedback? In an amplifier design where we begin with a 180° phase shift between the output and the input, it occurs at that point where the output signal, because of various additional phase delays in the system, is shifted another 180° (which puts it back in phase with the input signal). Actually, instability begins to occur before the 180° phase is reached. A safety margin of 40° (henceforth referred to as the *phase margin**) limits the acceptable phase shift through the system to less than 140°.

2-3 PHASE SHIFT AS A FUNCTION OF TIME DELAY

Since phase shift plays such an important part in a feedback system, it is desirable that we review phase shift in terms of time delay and frequency in terms of time. In any system, a delay in response due to transit time and physical inertia is expected. This delay results in an anticipated phase lag or time delay between the command input and output reaction. The amount of delay is closely related to the frequency—more specifically, the wave shape of the signal being amplified.

EXAMPLE 2-3

In Figure 2-11, if the period $T_1 = 1$ s and is the time of one cycle or 360°, a delay of 0.1 s must correspond to a lag of $\frac{1}{10}$ of 360°, or 36°.

Similarly, if $T_2 = 0.2$ s, the frequency is $1/T_2$ or 5 Hz; a delay

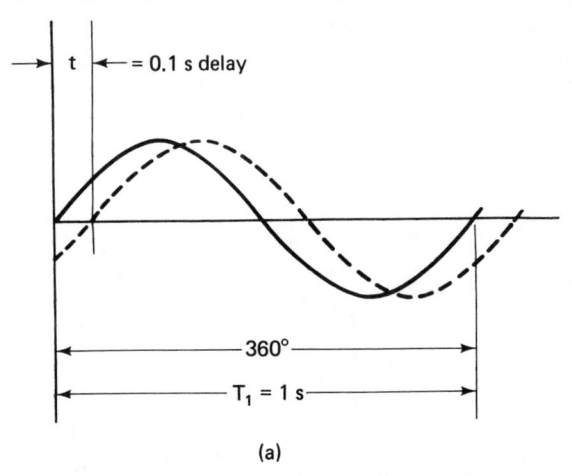

FIGURE 2-11 Delayed sine waves (phase and time relationship).

* Refer to Appendix D for a derivation of the 40° phase margin limitation.

(b)

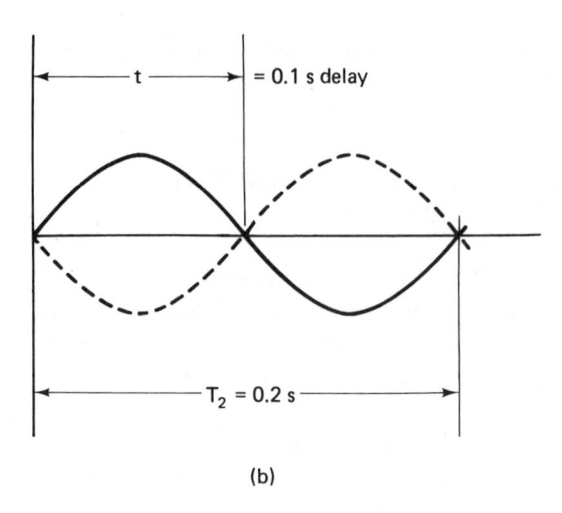

FIGURE 2-11 (Cont.).

of 0.1 s would now result in a phase lag of ½ of one cycle, or 180°. Put another way, in Figure 2-11(b), for each 180° of phase delay, the response lags by 0.1 s. In the same way, an 18° phase lag represents an 0.01-s or 10-ms delay in the output response, in which case phase lag is the same as time delay. The *RC*-series lag network provides just such a response. Since it closely resembles the mechanical response of a motor, the "heart" of an automatic control system, a detailed understanding of the simple *RC* lag network is required.

GLOSSARY

Beta: As it relates to feedback amplifiers—the percentage of the output signal fed back to the input of the system.

Block Diagram: The interconnection of subsystems that make up a complete system, showing the signal paths between the subsystems—sometimes also called a "flow diagram."

Isolation Amplifier: A noninverting amplifier with unity gain having a very high input impedance and a low output impedance, used primarily to prevent loading of a high-impedance driving source.

Lag Network: A combination of components connected so that the output voltage lags the applied (input) voltage; also referred to as a "delay network."

Negative Feedback: The addition of all or part of the output with the input signal to an amplifier or system, resulting in a decrease in the output response of the amplifier or system.

Positive Feedback: The addition of all or part of the output, with the input signal resulting in an increase in output response of an amplifier or control system. In most cases it will produce oscillation or some form of instability.

Summing Amplifier: An amplifier with a gain of unity with at least two inputs which are added to produce a single output; usually called "summing junction."

Transducer: Instruments that convert input energy of one form to another, usually mechanical to electrical. Examples of interest to our discussion are position potentiometers, which convert angular position to a voltage, and the tachometer, which converts revolutions per minute or radians per second to a voltage.

PROBLEMS

2.1 Determine the gain for each of the block diagrams.
(a) Given:

$v_i \longrightarrow \boxed{A} \longrightarrow v_o$ When A = 20

and

$v_i \longrightarrow \boxed{A_1} \longrightarrow \boxed{A_2} \longrightarrow v_o$

(b) When $A_1 = 10$ and $A_2 = -10$.
(c) When $A_2 = 10/\underline{-40°}$ and $A_1 = 10/\underline{20°}$.
(d) When $A_1 = +6$ dB and $A_2 = +8$ dB.
(e) When $A_1 = -6$ dB and $A_2 = +8$ dB.
(f) When $A_2 = -6$ dB and $A_1 = +6$ dB$/\underline{-180°}$.

2.2

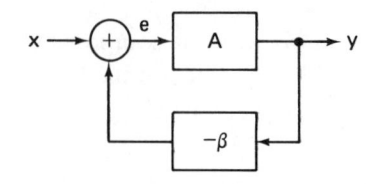

(a) Solve for e in terms of x, y, and β.
(b) Solve for y in terms of x, A, and β.
(c) Solve for TF or y/x when $A = 25$; $\beta = 0.04$.

2.3 Given:

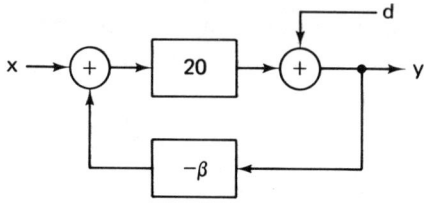

(a) Determine y in terms of x, d, A, and β where $A = 20$.
(b) If $x = 0.02$, $d = 0.02$, solve for y when $\beta = 0.1$.

2.4 Given:

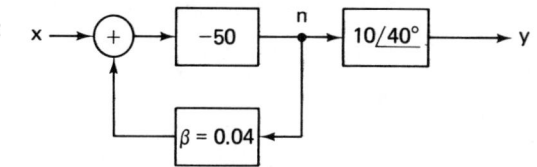

What is y when $x = 0.5$? (*Hint*: Solve for n.)

2.5 Given:

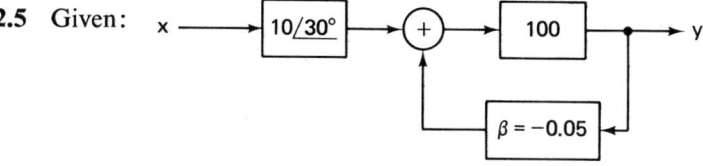

(a) What is the overall transfer function?
(b) What is the value of y when $x = 3$?

2.6 Given:

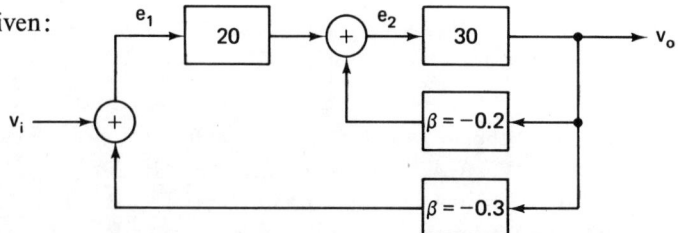

Solve for v_o with $v_i = 0.15$ V. (*Hint*: Solve for e_1, then for e_2.)

2.7 By observation, what is the approximate closed-loop gain of the following amplifiers?

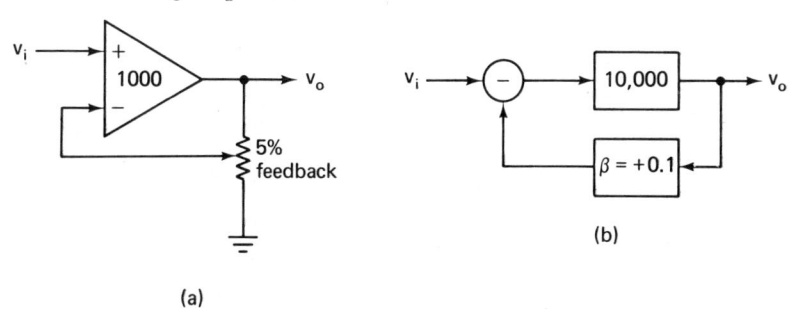

(a)

(b)

2.8 (a) Derive the gain equation for the amplifier in Figure 2-7(a).
(b) What happens to the gain of the equation found in Problem 2.8(a) when $\beta = 0°$?
(c) What happens to the closed-loop gain of part (b) when $A \cdot \beta = -1$?

2.9 What is the overall gain of the system shown? (*Note*: The second summing point subtracts.)

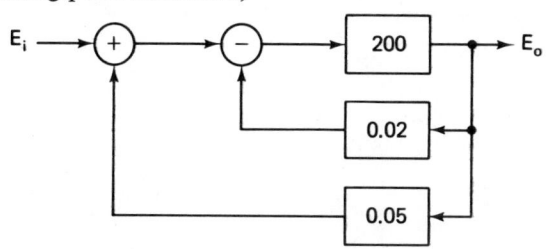

2.10 Determine v_o for the summing amplifier shown.

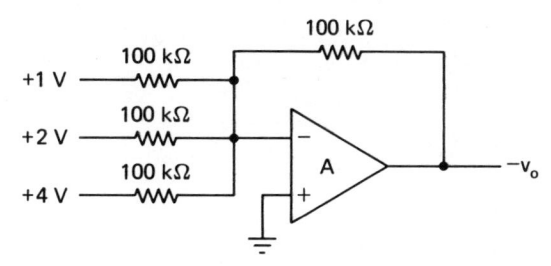

2.11 For the input voltage given, determine v_o.

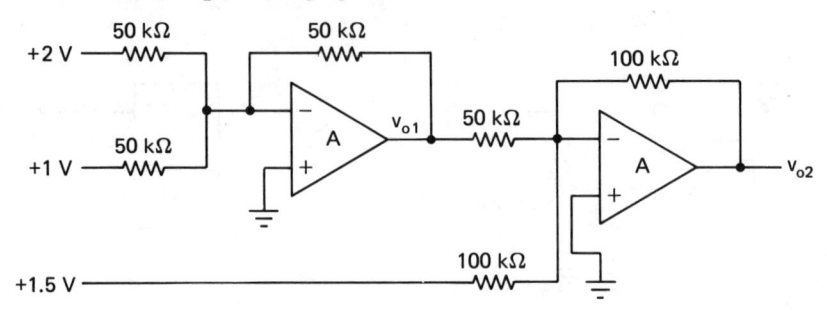

3

Graphical Solution of Network Response

3-1 INTRODUCTION

Closed-loop systems are made up of a number of blocks, each of which contains its own attenuation and phase characteristics, as shown in Figure 3-1. When *negative feedback* is introduced, care must be taken to assure that the sum of the phase shifts due to the phase-shift characteristic of each block does not approach 180° at a frequency where the *gain of the system is greater than 1*; otherwise, oscillation or instability may result (see Section 2-2).

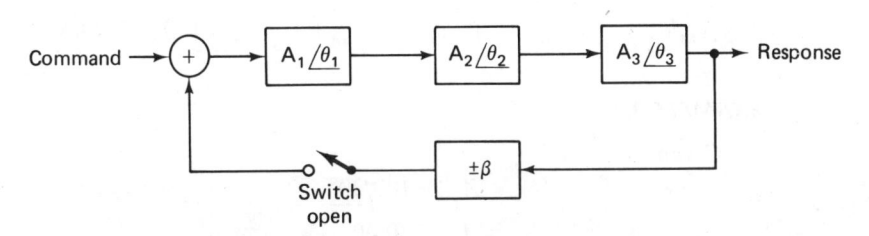

FIGURE 3-1 Flow diagram of a feedback system.

32   *Graphical Solution of Network Response*   *Ch. 3*

To help simplify the prediction of the overall system performance, that is, the addition of the amplitude-phase characteristics of all the subsystems, several important terms need to be mentioned. These include the decibel, the octave, the decade, the radian, and the corner frequency of a network or subsystem.

3-2 THE DECIBEL

It is customary to express the gain or attenuation characteristics of an amplifier or network in decibels rather than as simple numbers. The decibel (dB) is a logarithmic ratio and for voltage, current, and power is

$$\frac{v_o}{v_i} \text{(dB)} = 20 \log_{10} \frac{v_o}{v_i} \qquad (3\text{-}1)$$

$$\frac{i_o}{i_i} \text{(dB)} = 20 \log_{10} \frac{i_o}{i_i} \qquad (3\text{-}2)$$

and for power ratios,

$$\frac{P_o}{P_i} \text{(dB)} = 10 \log_{10} \frac{P_o}{P_i} \qquad (3\text{-}3)$$

The reason the decibel is such a valuable conversion unit is that multiplication becomes simply a matter of addition and division a matter of subtraction. When several blocks are combined as shown in the flow diagram of Figure 3-1, the overall open-loop gain is simply the sum of the individual amplifier gains in decibels.

The overall gain of Figure 3-1 is

$$A/\underline{\theta} = A_1 \cdot A_2 \cdot A_3 / \underline{\theta_1 + \theta_2 + \theta_3} \qquad (3\text{-}4)$$

However, if A_1, A_2, and A_3 are given in decibels, the overall gain in dB becomes simply

$$A_{\text{dB}}/\underline{\theta_T} = [A_{1(\text{dB})} + A_{2(\text{dB})} + A_{3(\text{dB})}]/\underline{\theta_1 + \theta_2 + \theta_3} \qquad (3\text{-}5)$$

EXAMPLE 3-1

Given

$$A_1 = 100/\underline{20°}$$
$$A_2 = 30/\underline{40°}$$
$$A_3 = 40/\underline{70°}$$

Sec. 3-3   The Time Constant and the Corner Frequency   33

SOLUTION:

In decibel form,

$$A_1 = 20 \log (100) = 40 \text{ dB}/\underline{20°}$$
$$A_2 = 20 \log (30) = 29.5 \text{ dB}/\underline{40°}$$
$$A_3 = 20 \log (40) = 32 \text{ dB}/\underline{70°}$$

so that

$$A_{dB}/\underline{\theta_T} = (40 + 29.5 + 32)/\underline{20° + 40° + 70°}$$
$$= 101.5 \text{ dB}/\underline{130°}$$

EXAMPLE 3-2

Given

$$A_1 = 50/\underline{30°}$$
$$A_2 = 0.5/\underline{-20°}$$
$$A_3 = 12/\underline{-90°}$$

SOLUTION:

$$A_1(\text{dB}) = 20 \log_{10} (50) = 33.98 \text{ dB}/\underline{30°}$$
$$A_2(\text{dB}) = 20 \log_{10} (0.5) = -6 \text{ dB}/\underline{-20°}$$
$$A_3(\text{dB}) = 20 \log_{10} (12) = 21.58 \text{ dB}/\underline{-90°}$$

so the overall gain becomes

$$A_{dB}/\underline{\theta_T} = (33.98 + 21.58 - 6)/\underline{30° - 20° - 90°}$$

and

$$A_{dB}/\underline{\theta_T} = 49.56 \text{ dB}/\underline{-80°}$$

3-3 THE TIME CONSTANT AND THE CORNER FREQUENCY

Let us begin with a review of the *sine wave*. If the period of a sine wave is 0.01 s, the frequency f in hertz is

$$f = \frac{1}{t} = 1/\text{s} = \frac{1}{0.01} = 100 \text{ Hz} \qquad (3\text{-}6)$$

Recall that one cycle = 360° and that 1 rad = 360°/2π = 57.3° or that there are 2π rad/cycle. Now when ω in rad/s is used to define frequency,

$$\text{frequency} = \frac{1}{\text{seconds}} = \omega = \frac{1}{\tau} = \text{rad/s} \qquad (3\text{-}7)$$

Note that τ is tau, not t, which is somewhat different. Recall that τ represents the time it takes the output (whether it is a voltage or the speed of a motor) to reach 63% of its final value.

This critical frequency, equal to $1/\tau$, is called the *corner frequency* ω_c. The series resistor–capacitor network of Figure 3-2 provides an excellent example of the importance of the time constant (where $\tau = RC$). The conclusions reached will apply to any

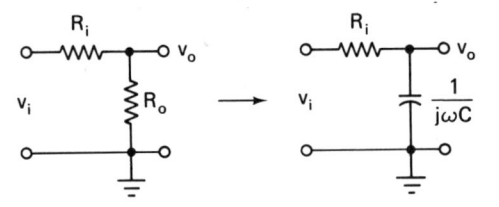

FIGURE 3-2 Voltage divider.

system where $1/\tau = \omega_c =$ rad/s. Referring to Figure 3-2 (using the voltage-divider rule),

$$\frac{v_o}{v_i} = \frac{R_o}{R_i + R_o} \qquad (3\text{-}8)$$

In a like manner, replacing R_o with the capacity C, whose capacitive reactance is $1/j\omega C$,

$$\frac{v_o}{v_i} = \frac{1/j\omega C}{R + 1/j\omega C} \cdot \frac{j\omega C}{j\omega C} = \frac{1}{j\omega RC + 1} \qquad (3\text{-}9)$$

Since $RC = \tau$, equation (3-9) can be rewritten as

$$\text{TF} = \frac{v_o}{v_i} = \frac{1}{1 + j\omega\tau} \qquad (3\text{-}10)$$

The graphical presentation of $(1 + j\omega\tau)$ of equation (3-10) is given in Figure 3-3. Equation (3-10) in polar form becomes

$$\frac{v_o}{v_i} = A\underline{/\theta} = \frac{1}{[1 + (\omega\tau)^2]^{1/2}}\underline{/-\tan^{-1}\omega\tau} \qquad (3\text{-}11)$$

Sec. 3-3   The Time Constant and the Corner Frequency   35

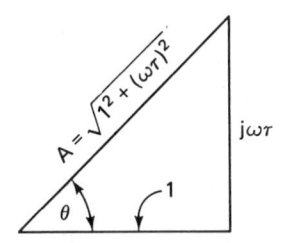

Where $A = \sqrt{1^2 + (\omega\tau)^2}$
$\theta = \tan^{-1}(\omega\tau)$

FIGURE 3-3 Graphical presentation of $1 + j\omega\tau$.

EXAMPLE 3-3

A network has a time constant $\tau = 0.01$ s. What is the corner frequency in rad/s? Also determine the corner frequency in hertz.

SOLUTION:

$$\omega_c = \frac{1}{\tau} = \frac{1}{0.01} = 100 \text{ rad/s}$$

the same as $100/2\pi$, or 16.9 Hz.

EXAMPLE 3-4

Given a series lag network with $\tau = 1$ s, calculate various values of the output/input voltage as the input frequency is varied.

Solutions of the gain $A/\underline{\theta}$ with $\tau = 1$ s for $\omega = 1$ and $\omega = 2$ rad/s are as follows: for $\omega = 1$,

$$A/\underline{\theta} = \frac{/-\tan^{-1}(1)}{\sqrt{1+1}} = 0.707/\underline{-45°} = -3 \text{ dB}/\underline{-45°}$$

and for $\omega = 2$,

$$A/\underline{\theta} = \frac{/-\tan^{-1}(2)}{\sqrt{1^2 + 2^2}} = \frac{/-63.4°}{\sqrt{5}} = -6.9 \text{ dB}/\underline{-63.4°} \quad (3.12)$$

Table 3-1 summarizes the amplitude and phase of the lag network with $\tau = 1$ s for a number of frequencies above and below 1 rad/s. These data will be used later to develop the univer-

TABLE 3-1 Amplitude and phase of the output of a delay network with a corner frequency of $\omega_c = 1$ rad/s.

ω	$\theta°$	$A\ (v_o/v_i)$	$A\ (dB)$
0.1	−5.7°	0.995	−0.04
0.2	−11.3°	0.962	−0.34
0.4	−21.8°	0.928	−0.64
0.6	−31°	0.857	−1.34
1.0	−45°	0.707	−3
2	−63.4°	0.45	−7
4	−76°	0.243	−12.3
6	−80.5°	0.164	−15.7
10	−84.3°	0.10	−20
20	−87.1°	0.05	−26
40	−88.6°	0.025	−32
100	−89.4°	0.01	−40

sal response curve of Figure 3-7. A plot of the data of Table 3-1 is presented in Figure 3-4. To understand the format used in the plot, an understanding of the octave and decibel is necessary.

3-4 THE OCTAVE AND THE DECADE

A dictionary definition of *octave* is "the difference between two frequencies, one of which has twice as many oscillations per second as the other." For example, the first octave above 400 H is 800 Hz, the second octave becomes 1600 Hz, the third 3200 Hz, and so on. However, in a like manner, the first octave above 1600 Hz would be 3200 Hz and the first above 3200 Hz would be 6400 Hz. The capacitive reactance is given as

$$-jX_c = \frac{1}{\omega C} \cdot \underline{/-90°} \quad \text{ohms} \quad (3\text{-}13)$$

Each time ω is doubled, *or moved up one octave*, the reactance is halved. For the series *RC* circuit of Figure 3-2, the output voltage obtained as a function of frequency shows that a few octaves beyond the corner frequency, each time the input frequency moves up one octave, the output voltage drops by a factor of 2, or by 6 dB.

Therefore, it is said that beyond the corner frequency, as the frequency is doubled (increased by an octave), the output voltage will be cut in half (reduced by 6 dB), or "the output of a single lag network decreases at the rate of 6 dB/octave."

The *decade*, by definition, is the difference between two frequencies, one of which has *10 times* as many oscillations per second as the other. For example, one decade above 400 Hz is 4000 Hz, the second decade above 400 Hz is 40,000 Hz, and so on. In a like manner, the first decade above 400 Hz is 4000 Hz and the first decade above 4000 Hz is 40,000 Hz. Again, a look at the data of Table 3-1 shows that several octaves beyond the corner frequency, for each increase in frequency of 10 times (one decade), the output decreases by a factor of 10 (-20 dB) and the output response is described as "having a slope of -20 dB/decade beyond the corner frequency."

The foregoing discussion can be shown clearly by using the data of Table 3-1 to make a plot (Figure 3-4) of gain (A) in dB versus frequency in rad/s. The amplitude (A) is a linear plot on the

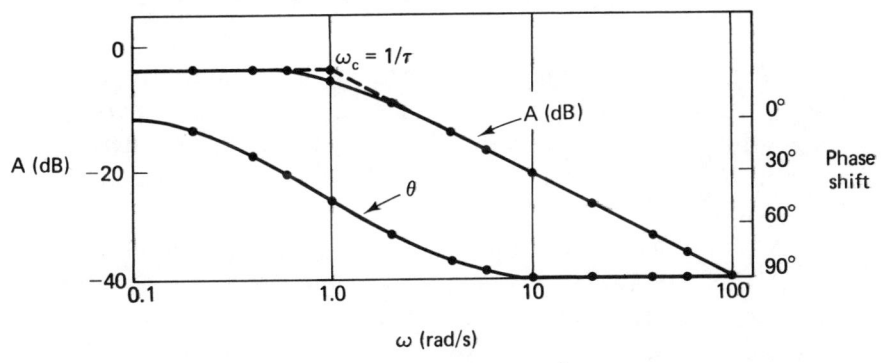

FIGURE 3-4 Plot of Table 3-1: $A\underline{/\theta} = \dfrac{1}{\sqrt{1+\omega^2}} \underline{/-\tan^{-1}(\omega)}$.

y axis (recall that dB is a logarithmic ratio), and the frequency is plotted on a logarithm scale along the x axis in multiples of 10 to correspond to the decade (i.e., 1, 10, 100, 1000). Therefore, as ω increases logarithmically, A, which is in dB, decreases linearly and the straight-line amplitude response is obtained beyond the corner frequency $\omega_c = 1/\tau$. For these reasons, semilog paper will be used for all amplitude and phase plots in the remainder of the text.

3-5 THE UNIVERSAL PLOT OF A SINGLE LAG NETWORK

We showed in Section 3-4 that if a corner frequency of $\omega_c = 1$ is used, then based on the 6 dB/octave or 20 dB/decade slope that results beyond $\omega_c = 1$, a table can be prepared for multiples of ω_c versus amplitude in decibels (Table 3-2). A plot of this table becomes the graph of Figure 3-5.

TABLE 3-2 Amplitude as ω increases (linearized data).

$\omega_1 = \omega_c$	A (dB)
ω_1	≈ 0
$2\omega_1$	≈ -6
$4\omega_1$	≈ -12
$10\omega_1$	≈ -20
$100\omega_1$	≈ -40
$1000\omega_1$	≈ -60

When greater accuracy is required than that obtained from the straight-line approach (as will be called for in Chapter 9), applying a -3-dB correction at the corner frequency as shown by the dashed line in Figure 3-5, results in very accurate amplitude data at all frequencies.

A plot of θ versus ω for a single delay network with $\omega_c = 1$ rad/s (data of Table 3-1) is shown in Figure 3-6. As was done for the amplitude A, a very accurate graphical technique for determining the phase shift (θ) is used. At the corner frequency (ω_c), we know that θ is $-45°$. Locate this point on your frequency-phase plot. One decade below ω_c, let $\theta = 0°$ and at 1 decade above ω_c, let $\theta = -90°$. Then draw a straight line through these points. The straight-line plot of Figure 3-6 provides very close approximations of θ versus ω; the error at any point along the line does not exceed 5.7°.

Since the margin of safety in our designs, referred to as the *phase margin*, is to be at least 40°, the results obtained by the straight-line method used in Figures 3-5 and 3-6 are adequate for most of our system evaluations. Figure 3-7 summarizes the results

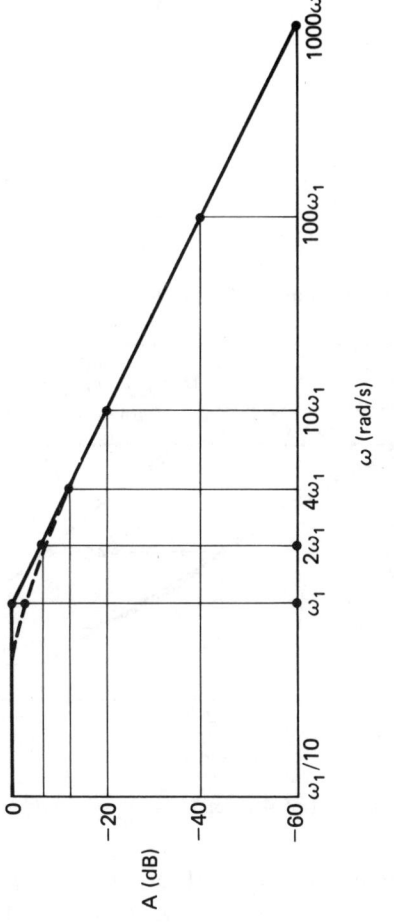

FIGURE 3-5 Plot of Table 3-2.

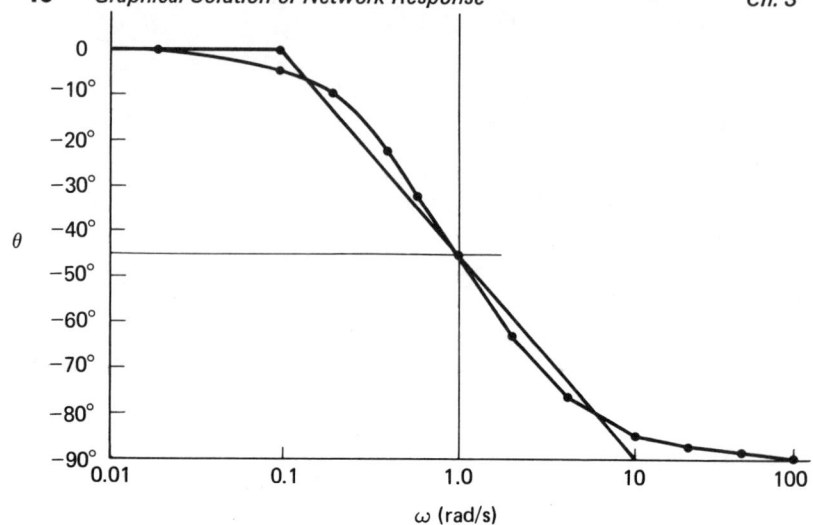

FIGURE 3-6 Plot of phase angle of a single lag network (data of Table 3-1).

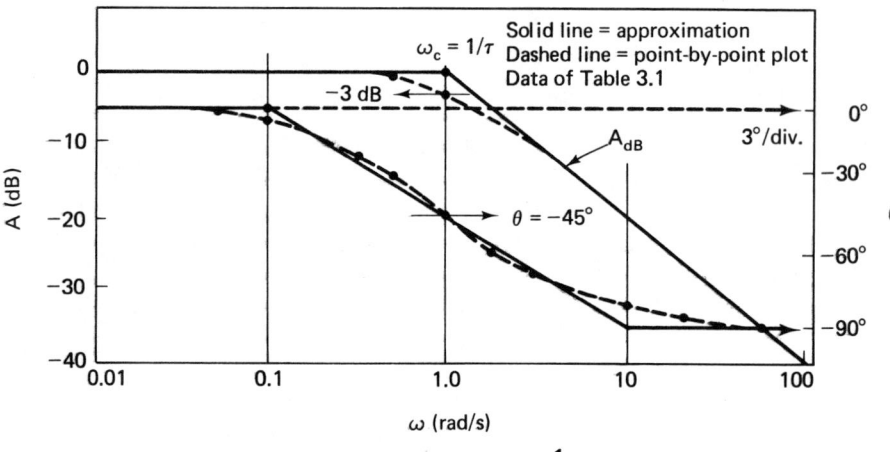

FIGURE 3-7 Plot of $A/\underline{\theta}$ for TF = $\dfrac{1}{j\omega\tau + 1}$ with $\tau = 1$.

described in the analysis above. It provides a *universal* plot of amplitude in decibels and phase shift in degrees for a delay network with one corner frequency. This procedure will be extended to cover several networks in series.

3-5-1 Normalizing

To use the universal plot of $A/\underline{\theta}$ for solutions of networks with corner frequencies other than $\omega_c = 1$ requires the determination of a multiplier constant called *n*. Simply, if a frequency, say

Sec. 3-5   The Universal Plot of a Single Lag Network   41

50 rad/s is the corner frequency of a network, the plot of this network remains that of Figure 3-7 except that ω along the x axis must be multiplied by 50. Figure 3-8 will be used to show how the normalizing constant is used to solve for the network response at any frequency of a lag network with $\tau = 0.02$ s.

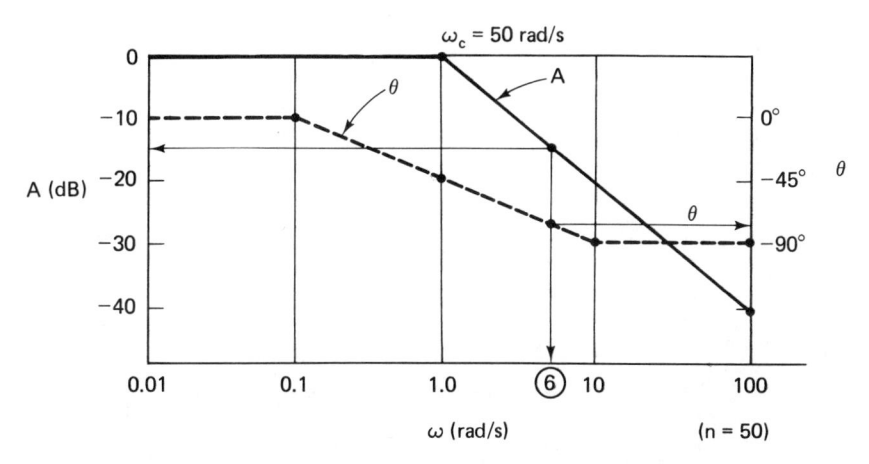

FIGURE 3-8 Solution of Example 3-1 where $n = 50$ and $\omega = 6n = 300$ rad/s.

EXAMPLE 3-5

Given the lag network with $\tau = 0.02$ s, what is the amplitude and phase of the output voltage at 300 rad/s?

SOLUTION:

1. Note on the linearized graph of Figure 3-8 that the frequency multiplier $n = 50$.

2. 300 rad normalized to $n = 50$ is $300/50 = 6n$.

3. At 6 on the normalized plot, draw a vertical straight line through the amplitude and phase curves.

4. Read the amplitude and phase for 300 rad/s at the normalized frequency $\omega_n = 6$, $A = -16$ dB and $\theta = -82°$.

Figure 3-8 shows that at 300 rad/s, the output voltage for the network of example 3-5 will be down -16 dB and the phase shift between the output voltage and input voltage is 82° lagging ($-82°$), so that $A/\theta = -16$ dB$/-82°$.

EXAMPLE 3-6

Using the normalized plots of A and θ for $\omega_c = 1$, determine the attenuation and phase shift of a lag network with $\tau = 0.05$ s ($\omega_c = 20$ rad/s) at a frequency that is 2 octaves above its corner frequency.

SOLUTION:

STEP 1:

Two octaves above $\omega_c = 20$ rad/s is determined as follows:

1. $2\omega_c = 2 \times 20 = 40$ rad/s = first octave above 20 rad/s.
2. $2 \times$ first octave = $2 \times 40 = 80$ rad/s = second octave above 20 rad/s.
3. With $n = 20$, 80 rad is normalized to $80/20 = 4$, as shown in Figure 3-9.

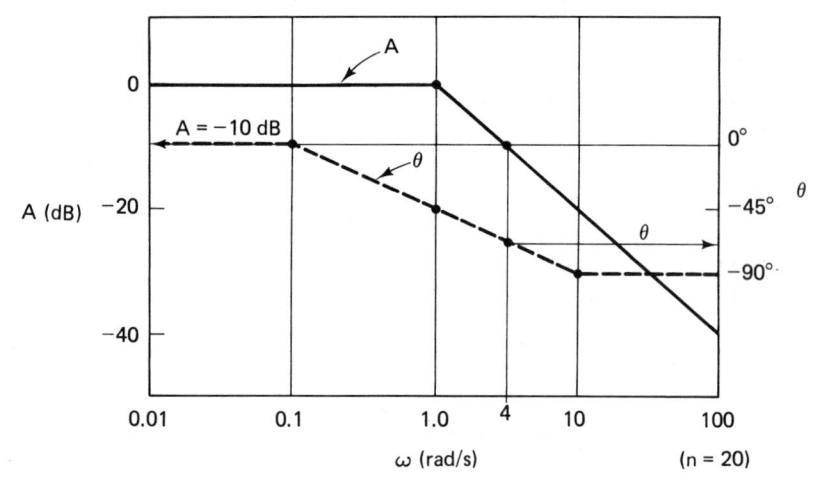

FIGURE 3-9 Determining $A/\underline{\theta}$ at $4n = 80$ rad/s.

STEP 2:

Following the procedures described in Example 3-5, draw a vertical line at $\omega = 4$ on the normalized plot of Figure 3-9.

STEP 3:

It can be seen that the output responses of the network two octaves above the corner frequency $\omega_c = 20$ rad/s or at $\omega = 80$ rad/s is down approximately -10 dB with $\theta = -72°$.

3-6 SOLUTIONS OF NETWORKS IN SERIES

3-6-1 The Bode Plot

The graphical solution of the response of several networks in series will be done by the addition of the response of each individual network on a common plot with the amplitude in dB on the y axis and the frequency in rad/s on the x axis (see the universal plot of Figure 3-7). This type of graphical presentation has come to be known as the *Bode plot* or Bode diagram.* To demonstrate the general technique for the graphical solution of multiple networks using the Bode plot, several practical problems will be solved, the main objective being:

1. To become more familiar with the use of normalized curves.
2. To construct and analyze Bode plots using more than one corner frequency.
3. To solve the overall response characteristics of several networks in series.

A simple procedure for plotting the frequency characteristics of two or more networks in series is to plot each one separately and then add them graphically to provide the overall response (adding dB being the same as multiplying).

EXAMPLE 3-7

Let us take two networks, such as T_A and T_B, and add their outputs. (Note that an amplifier, called an isolation amplifier, is placed between the two networks to reduce the loading effects between them.) Loading could alter the overall frequency response considerably.

SOLUTION:

In the series network of Figure 3-10, because the *lowest corner frequency* will be the first to affect the output in a multiple-network system as the frequency is increased from zero, the Bode plot will be "normalized" to the *lowest corner frequency* in the

* Named after H. W. Bode of the Bell Telephone Laboratories.

44   *Graphical Solution of Network Response*   *Ch. 3*

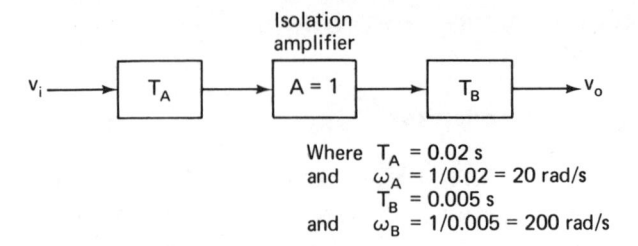

$$v_i \longrightarrow \boxed{T_A} \longrightarrow \boxed{\substack{\text{Isolation} \\ \text{amplifier} \\ A = 1}} \longrightarrow \boxed{T_B} \longrightarrow v_o$$

Where $T_A = 0.02$ s
and $\omega_A = 1/0.02 = 20$ rad/s
$T_B = 0.005$ s
and $\omega_B = 1/0.005 = 200$ rad/s

FIGURE 3-10 Two networks in series with an isolation amplifier.

system. Thus, for the networks under consideration, $n = 50$ rad/s will govern n, and n becomes 50. ω_B normalized to $n = 50$ becomes $200/50 = 4n$. Actually, the plot of each network is made separately and the output and phase of each then added, resulting in the overall response of the two networks being the sum of Figure 3-11(a) and (b), shown as curve (c) in Figure 3-11.

Before analyzing the meaning of the overall response of Figure 3-11(c), it will be recalled that amplification must also be present before instability can become a problem. (i.e., the gain is greater than 1, or 0 dB).

3-6-2 The Amplifier

The presentation of amplification is quite straightforward. Instead of using A to represent the output/input ratio, it is replaced by K in decibels. Later K will be modified by its frequency characteristic (G) to KG. For the present, any gain A must be converted to gain in decibels for Bode plot presentation. Thus, for $A = 10$,

$$K = 20 \log_{10}(10) = +20 \text{ dB} \tag{3-14}$$

For $A = 100$,

$$K = +40 \text{ dB}$$

for $A = 0.1$,

$$K = -20 \text{ dB}$$

and for $A = 0.01$,

$$K = -40 \text{ dB}$$

On a Bode plot, K is simply a straight line in dB with $\theta = 0°$.

Sec. 3-6 Solutions of Networks in Series **45**

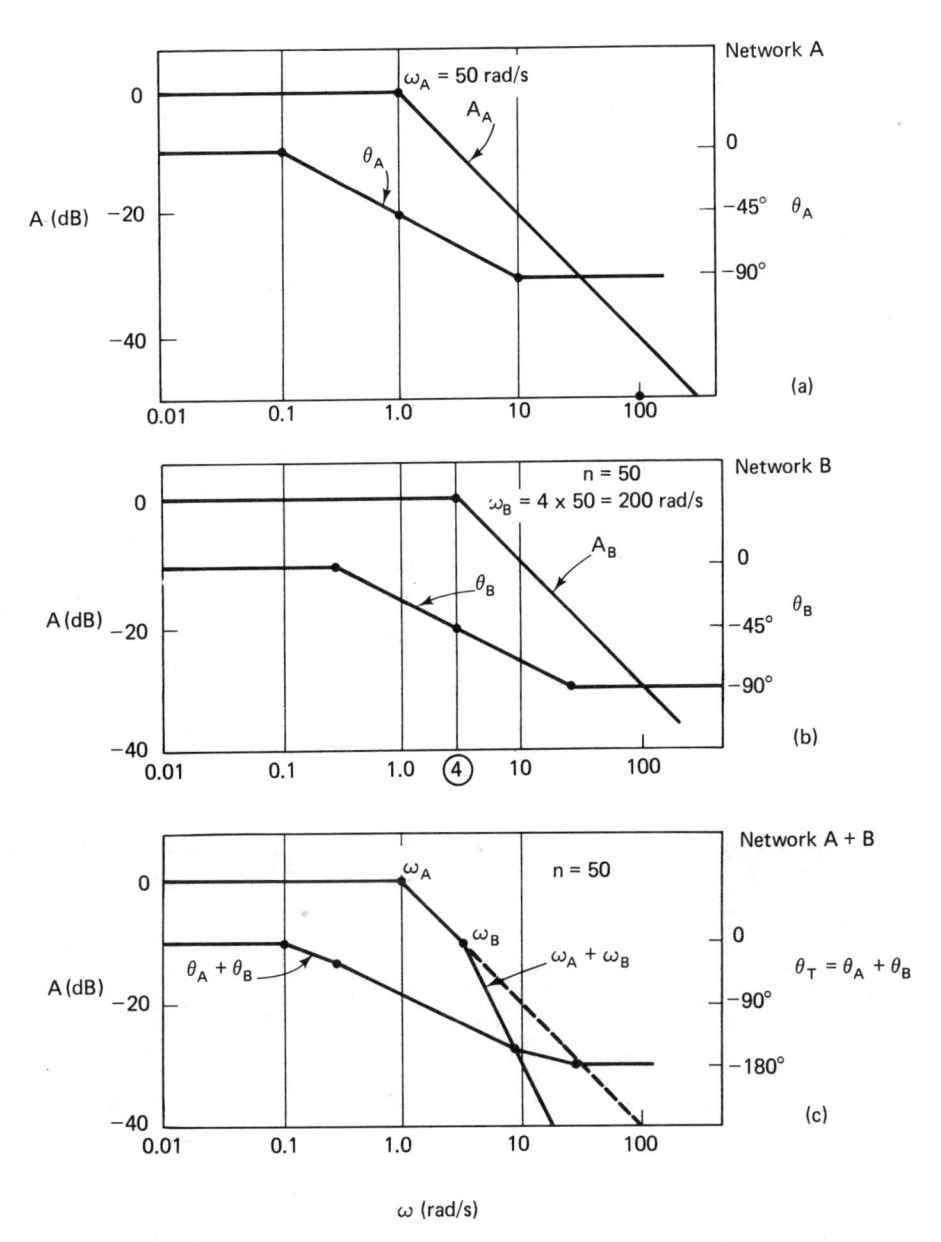

FIGURE 3-11 Adding of two networks by summing the Bode plots.

Several plots of K are shown in Figure 3-12. Adding an amplifier to Figure 3-10 would modify the block diagram to that of Figure 3-13.

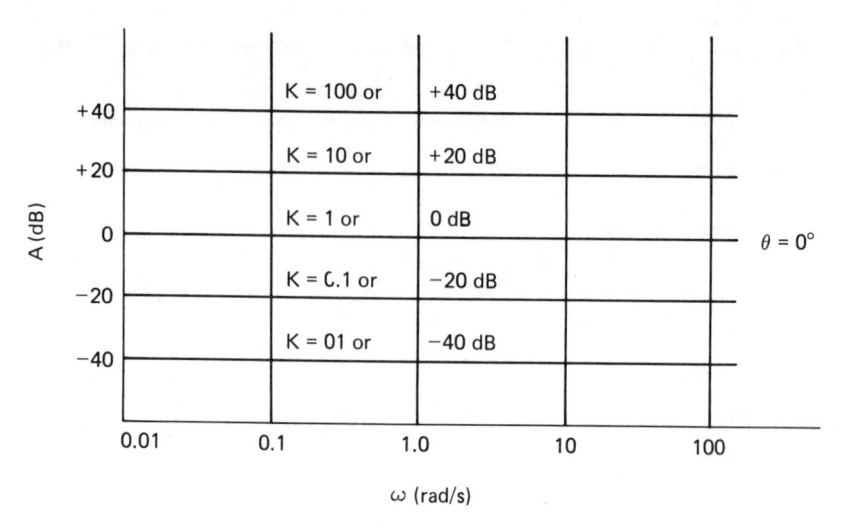

FIGURE 3-12 Bode plot of an amplifier with gain A in terms of K.

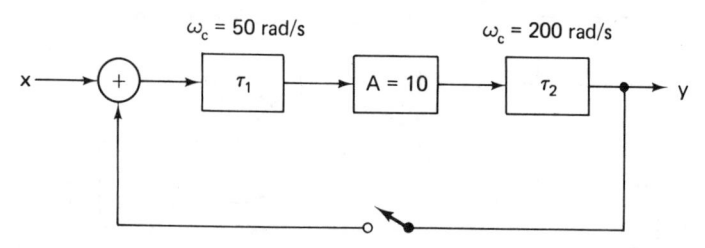

FIGURE 3-13 Amplifier, including several networks in series.

In Example 3-7, if the amplifier A is added and $A = 10$, ($K = +20$ dB), the new plot is the same as Figure 3-11(c) except that 0 dB becomes 20 dB (0 + 20 dB), as shown in Figure 3-14:

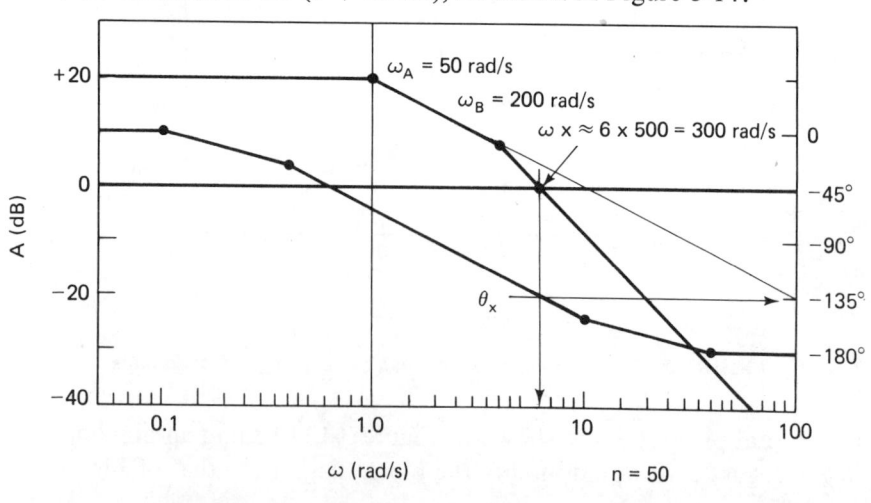

FIGURE 3-14 Adding gain of $K = +20$db to Figure 3-11(c).

46

3-6-3 Gain Crossover Frequency

We have mentioned that in a closed-loop system, when the phase shift approaches 180° and the gain is greater than 1 (0 dB), the system will become increasingly unstable, which would necessitate some form of phase correction. The frequency at which the amplitude response curve crosses the 0-dB line is known as the *gain crossover frequency* and is very important in analyzing system performance.

In Figure 3-14 this frequency is shown as $\omega_x \approx 300$ rad/s. At this frequency, phase is measured and the difference between θ and $-180°$ is called the *phase margin* of the system. It is generally accepted that at least a 40°* phase margin $(180° - \theta)$ is required for an acceptable design. In this example, $\theta = -135°$ and the phase margin is $180° - 135° = 45°$. Since ϕ, the phase margin, is greater than 40°, it is assumed the system will be stable when the feedback loop is closed.

For more accurate phase-shift data at the gain crossover frequency (ω_x), Table 3-3 shows the actual phase shift for various ratios of ω_x/ω_c, where ω_c is the lowest corner frequency in the system.

TABLE 3-3 Phase shift for various values of ω_x/ω_c.

$\tan^{-1}\left(\dfrac{\omega_x}{\omega_c}\right)$	Actual $\theta°$	Approx. $\theta°$	Error (°)
0.01	−0.59	0	−0.59
0.05	−2.9	0	−2.9
0.1	−5.7	0	−5.7
0.4	−21.8	−27	+5.2
0.8	−38.7	−40.8	+1.8
1.0	−45	−45	0
1.5	−56.3	−53	−3.3
2	−63.5	−58.5	−5
4	−76	−74	−2
6	−80.5	−80	−0.5
8	−82.8	−83	+0.2
10	−84.3	−90	+5.7
20	−87.2	−90	+2.8
100	−89.5	−90	+0.5

* See Appendix D.

EXAMPLE 3-8

In Figure 3-14, $\omega_x \approx 300$ rad/s for $\omega_A = 50$ rad/s and $\omega_B = 200$ rad/s. Determine the exact phase shift from the data of Table 3-3.

SOLUTION:

For $\omega_x/\omega_A = 300/50 = 6$,

$$\text{phase shift} = -80.5°$$

For $\omega_x/\omega_B = 300/200 = 1.5$,

$$\text{phase shift} = -56.3°$$
$$\overline{\text{total } \theta \quad = -136.8°}$$

Thus, the actual phase shift is $-136.8°$ instead of the $-134°$ obtained from the graph. The error of only a few degrees is insignificant. Since the error for any single network will not exceed $\pm 5.7°$, as can be seen in Table 3-3, the straight-line method for approximating phase shift is adequate.

Summary

Some very interesting observations can be made from the data and plots developed so far:

1. Most control systems take time to respond to a command (time = delay = phase shift = $-\theta$).

2. Using the decibel to describe gain is important because adding gain in decibels is the same as multiplying (a gain of $1 = 0$ dB).

3. Corner frequency $= \omega_c = $ rad/s $= 1/\tau$.

4. The slope of the amplitude line shows it to be approximately 6 dB/octave or 20 dB/decade beyond the corner frequency for each network used.

5. The phase shift (θ_0) versus frequency on a Bode plot very nearly approaches a straight line above and below ω_c. At ω_c, (the corner frequency) $\theta = -45°$; one octave above and

one octave below ω_c, the phase shift is approximately 0° and −90°, respectively.

6. By making slight corrections, very accurate amplitude and phase data can be obtained by the use of straight-line Bode plots.
7. Normalizing permits the use of a universal graph of $A/\underline{\theta}$ for quick solutions of networks with corner frequencies other than $\omega_c = 1$ by use of the normalizing constant n.
8. Amplification is accomplished graphically by shifting the universal plot up or down by the amount of gain desired in decibels.

GLOSSARY

Bode Plot: A semilog plot with the x axis representing frequency (ω) on a logarithmic scale and the y axis representing the output/input ratio in decibels on a linear scale.

Corner Frequency: A critical frequency specifically defined by $1/\tau = \omega_c$, whose output $A/\underline{\theta} = -3$ dB$/\underline{-45°}$.

Decade: The interval between two frequencies, one of which is 10 times the frequency of the other.

Gain Crossover Frequency: The upper frequency response, in a system with positive gain, where the gain has decreased to 0 dB or 1.

Normalize: The technique of specifying the lowest corner frequency in a system as $\omega_c = 1$ multiplied by the normalizing constant n, then to reference all other frequencies in terms of the multiplier n.

Octave: The interval between two frequencies, one of which is twice the frequency of the other.

Phase Margin: The difference between the actual phase shift at the gain crossover frequency and 180°.

PROBLEMS

3.1 Given:

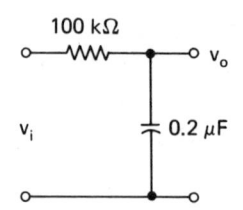

[Circuit: V_i — 100 kΩ resistor — V_o, with 0.2 μF capacitor to ground]

What is the amplitude and phase of the output three octaves above the corner frequency? (Use a normalized Bode plot to find the answer with $n = 50$).

3.2 Given:

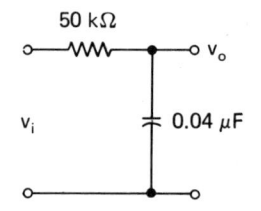

[Circuit: V_i — 50 kΩ resistor — V_o, with 0.04 μF capacitor to ground]

(a) What is the response four octaves above the corner frequency? Use the plot made for Problem 3.1 with $n = 500$.

(b) Prove the graphical solution mathematically for this frequency substituting into

$$A\underline{/\theta} = \frac{1\underline{/-\tan^{-1}(\omega\tau)}}{j\omega RC + 1}$$

(c) What is the output of the network at $\omega = 1/\tau$ rad/s?

3.3 (a) If two lag networks are cascaded

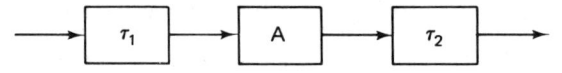

[Block diagram: τ_1 → A → τ_2]

with $A = 1$, $\tau_1 = 0.1$ s and $\tau_2 = 0.5$ s, draw the Bode plot showing the amplitude and phase.

(b) What is the response at 20 rad?

(c) What frequency does 20 rad/s represent?

3.4 Given:

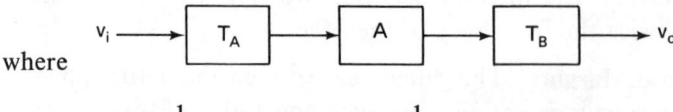

[Block diagram: V_i → T_A → A → T_B → V_o]

where

$$T_A = \frac{1}{j\omega(0.05) + 1}, \quad T_B = \frac{1}{j\omega(0.2) + 1}, \quad \text{and} \quad A = 100$$

(a) Develop the Bode plot showing amplitude and phase.
(b) What is the gain crossover frequency?
(c) What is the phase margin?
(d) What is the response two octaves beyond the gain crossover frequency?
(e) What is the output at 50 rad/s (both A and θ)? Prove your answer mathematically.

3.5 Given:

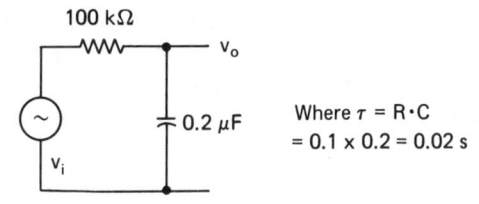

[Circuit: v_i source, 100 kΩ resistor, 0.2 μF capacitor, output v_o]

Where $\tau = R \cdot C$
$= 0.1 \times 0.2 = 0.02$ s

(a) Show mathematically that when $R = X_c$, $\omega_c = 1/\tau = 50$ rad/s.
(b) That at 200 rad/s, the output is down -12 dB, to $\frac{1}{4}$ or $0.25 v_i$.

3.6 Refer to the Bode plot of Problem 3.4.
(a) Determine $A/\underline{\theta}$ at $\omega = $ three octaves above ω_B where $\tau_B = 0.2$ s.
(b) Prove your answer mathematically.
(c) What value of C must be used if $R = 10{,}000\ \Omega$ to get a $\tau = 0.2$ s?

3.7 Given two networks, where $\tau_1 = 0.1$ s and $\tau_2 = 0.04$ s, find A_T and θ_T at $\omega = 30$ rad/s by solving for $A_1/\underline{\theta_1}$ and $A_2/\underline{\theta_2}$ and adding the outputs.

3.8 Given TF $= K/(1 + j\omega\tau)$, determine $A/\underline{\theta}$ at $= 20$ rad/s with $K = 30$ and $\tau = 0.2$ s.

3.9 Given:

[Block diagram: input C into summing junction, then Amp gain 20, then Motor with transfer function $\dfrac{K_v}{J\omega(J\omega\tau_m + 1)}$, output R, with feedback]

This system includes an amplifier with a gain of 20 driving a servo motor with a transfer function equal to $K_v/j\omega\,(j\omega\tau_m + 1)$, where $K_v = 5$ rad/s/V and $\tau_m = 0.5$ s. The complete transfer function becomes

$$\text{TF} = \frac{(20)(5)}{j\omega(0.5j\omega + 1)} = \frac{100}{j\omega(0.5j\omega + 1)}$$

(a) What is the amplitude and phase of the output signal at the corner frequency for this system? Solve by substituting ω_c for ω in the equation given.

(b) What is the amplitude and phase of the output one octave above ω_c?

(c) What is the amplitude and phase of the output one decade above ω_c?

3.10 (a) In the system shown in Problem 3.9, what happens to the phase shift of the output at the corner frequency if the amplifier gain is reduced from 20 to 5?

(b) What happens to the phase of the output one decade above ω_c with the reduced gain?

(c) What is your conclusion regarding the effect of gain changes on the phase shift of the output at any particular frequency?

4

The *s* Operator and the Laplace Transform

4-1 INTRODUCTION

At this point in our study of open- and closed-loop electrical systems, we could proceed to a rather complete analysis of a practical automatic control system. The open-loop response varies little from the Bode plot responses we made of several lag networks and amplifiers in cascade.

In the evaluation of practical systems, the phase margin at the gain crossover frequency must be established and when it is not within prescribed limits, phase compensation must be added. The addition of phase compensation and the analysis of the *closed-loop response* under steady state or a step input results in some very complex equations.

In the frequency domain, it is basically a problem involving complex algebra ($R \pm jX$) with numerous solutions, one for each frequency needed to produce a plot of the overall system response. In the time domain, as the systems become more complicated, the solutions for the output response to a step input as the command signal tend to become lengthy and time-consuming. The use of special mathematical techniques, the *s operator* and the *Laplace transforms*, simplifies and speeds up the solution of these cumbersome equations to provide system-response solutions in terms of both frequency and time.

4-2 BASIC TYPES OF NETWORKS

Networks play a very important role in system design: first, because they are used to compensate for excessive phase shift in a system; and second, because the equations that describe the output response of a network are of the same mathematical form that describe the behavior of the mechanical subsystems that make up a complete closed-loop control system. There are four basic types of networks associated with automatic control systems:

1. Resistive divider.
2. Lag network.
3. Lead network.
4. Series resonant circuit.

4-2-1 The Lag Network

The voltage divider has been adequately described and for Figure 4-1(a),

$$\frac{v_o}{v_i} = \frac{R_o}{R_i + R_o} \tag{4-1}$$

Similarly, the lag networks of Figure 4-1(b) were shown to be

$$\frac{v_o}{v_i} = A/\underline{\theta} = \frac{1}{[1 + (\omega\tau)^2]^{1/2}} /\underline{-\tan^{-1}(\omega\tau)} \tag{4-2}$$

4-2-2 The Lead Network

The third is a special network similar to the network above but provides a transfer function with a leading phase angle. It is used in various forms to modify the system response by correcting or "compensating" for too much lag in a system. Simply interchang-

Sec. 4-2   Basic Types of Networks   55

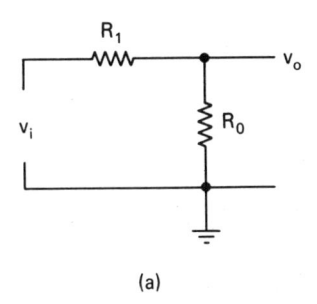

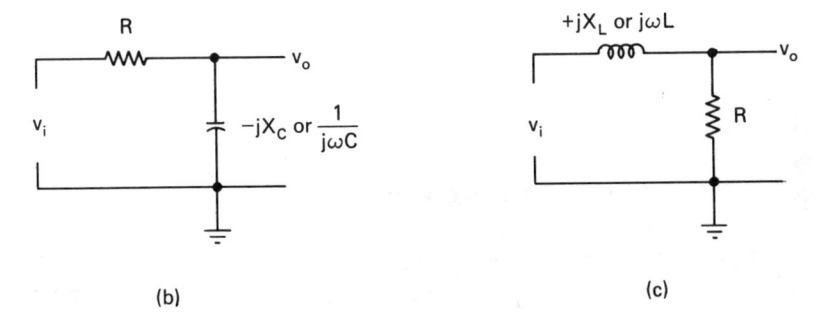

FIGURE 4-1 Voltage dividers and the lag network: (a) resistor voltage divider; (b) *RC* lag network; (c) *LR* lag network.

ing the resistors and reactances of Figure 4-1(b) will convert these networks into lead networks.

Referring to Figure 4-2 and using the voltage-divider rule for the resistive–capacitive circuit:

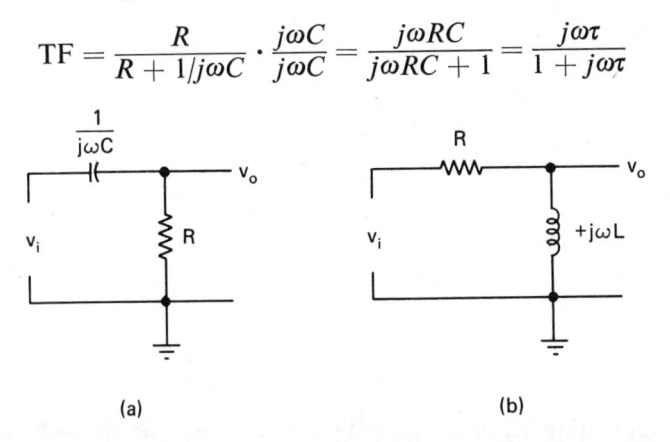

$$\text{TF} = \frac{R}{R + 1/j\omega C} \cdot \frac{j\omega C}{j\omega C} = \frac{j\omega RC}{j\omega RC + 1} = \frac{j\omega\tau}{1 + j\omega\tau}$$

FIGURE 4-2 Basic lead networks: (a) *RC* lead network; (b) *RL* lead network.

from which

$$A = \frac{\omega\tau}{[1 + (\omega\tau)^2]^{1/2}} \tag{4-3}$$

and

$$\theta = \underline{/+90° - \tan^{-1}(\omega\tau)} \quad \text{or} \quad \underline{/+\tan^{-1}\left(\frac{1}{\omega\tau}\right)} \tag{4-4}$$

Similarly, for the inductance–resistor circuit,

$$\text{TF} = \frac{j\omega L}{j\omega L + R} \cdot \frac{1/R}{1/R} = \frac{j\omega L/R}{j\omega L/R + 1} = \frac{j\omega\tau}{1 + j\omega\tau} \tag{4-5}$$

which provides the same results for $A/\underline{\theta}$ given by equations (4-3) and (4-4).

4-2-3 Miscellaneous Networks

There are a series of networks that are variations of the lead and lag networks and appear in many forms. A few of the common types are shown in Figure 4-3. For the lead network of Figure 4-3(a),

$$\text{TF} = \frac{R_2}{R_1 + R_2 + 1/j\omega C} \cdot \frac{j\omega C}{j\omega C} = \frac{j\omega R_2 C}{(R_1 + R_2)j\omega C + 1} \tag{4-6}$$

or

$$\text{TF} = \frac{j\omega\tau_1}{1 + j\omega\tau_2} \tag{4-7}$$

where $\tau_1 = R_2 C$
$\tau_2 = (R_1 + R_2)C$

from which

$$A = \frac{\omega\tau_1}{[1 + (\omega\tau_2)^2]^{1/2}} \tag{4-8}$$

and

$$\theta = \underline{/+90° - \tan^{-1}(\omega\tau_2)} \tag{4-9}$$

The remaining circuits are left to the reader to develop as an exercise in complex algebra.

Sec. 4-6   *Use of the s Operator and the Laplace Transforms*   **69**

This closely resembles pair 6, which is equal to

$$A\left[\frac{1}{s(s\tau + 1)}\right]$$

With $A = E/R$:

$$i(t) = A(1 - e^{-t/\tau}) = \frac{E}{R}(1 - e^{-t/\tau})$$

where $\tau = L/R$.

We should recognize the solution above as the buildup of current in a lead network.

EXAMPLE 4-11

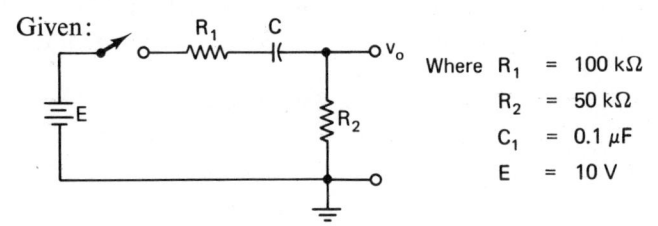

Given:

Where $R_1 = 100 \text{ k}\Omega$
$R_2 = 50 \text{ k}\Omega$
$C_1 = 0.1 \text{ }\mu\text{F}$
$E = 10 \text{ V}$

determine v_o at 0.01 s after closure of the switch.

SOLUTION:

$$I(s) = \frac{E(s)}{Z(s)} = \frac{10}{s} \cdot \frac{1}{100K + 50K + 10^6/0.1s}$$

$$= \frac{10}{s(150K + 10 \times 10^6/s)}$$

With R in megohms and C is microfarads,

$$I(s) = \frac{10}{s(0.15 + 10/s)} \quad \text{or} \quad \frac{10}{0.15s + 10} = \frac{66.7}{s + 66.7}$$

The solution readily fits into the form of pair 3, which is $A/(s + a)$ if we make $A = 66.7$ and $a = 66.7$. Then for $F(s) = A/(s + a)$,

$$f(t) = A \cdot e^{-at} \quad \text{and} \quad i(t) = 66.7e^{-66.7t}$$

Thus, for $t = 0.01$ s,

$$i = 66.7e^{-66.7(0.01)} = 66.7e^{-0.667}$$
$$= 66.7(0.51) = 34.2 \times 10^{-6} \text{ A}$$

so that

$$v_o = i \cdot R_2 = (34.2 \times 10^{-6})(0.05 \times 10^6) = 1.7 \text{ V}$$

Therefore, the output voltage of the lead network is 1.7 V 0.01 s after closing the switch.

4-7 THE SERIES *LCR* CIRCUIT

Some very interesting network solutions are possible using the *s* operator and the Laplace transforms. Take the familiar resonant circuit shown in Figure 4-8. It should be recognized that the

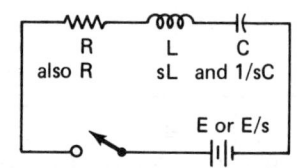

FIGURE 4-8 Series resonant circuit.

analysis of this circuit can be accomplished in terms of frequency, or, more in line with the Laplace technique, by using a step input (E/s or $1/s$). When a step input is used as the input signal, the analysis is referred to as the *transient response to a step input*. With a step voltage $E(s) = E/s$ applied to the system (by closing the switch to the battery),

$$I(s) = \frac{E(s)}{Z(s)} = \frac{E}{s} \frac{1}{sL + R + 1/sC} = \frac{E}{s^2 L + Rs + 1/C} \quad (4\text{-}13)$$

Dividing the numerator and denominator by L results in

$$I(s) = \frac{E}{L} \cdot \frac{1}{s^2 + R/L \cdot s + 1/LC} \quad (4\text{-}14)$$

A look at the table of transforms shows that pair 9 looks very much like (4-11):

$$F(s) = A \cdot \frac{1}{s^2 + 2as + \omega_n^2} \qquad (4\text{-}15)$$

which transforms to

$$f(t) = \frac{A}{\omega_d} \cdot e^{-at} \sin \omega_d \cdot t \qquad (4\text{-}16)$$

where $\omega_n = \sqrt{\omega_d^2 + a^2}$. Since equation (4-14) is equal to that of (4-15),

$$\frac{E}{L} \cdot \frac{1}{s^2 + R/L \cdot s + 1/LC} = A \cdot \frac{1}{s^2 + 2as + \omega_n^2} \qquad (4\text{-}17)$$

from which

$$\frac{E}{L} = A$$

$$\frac{R}{L} = 2a \quad \text{and} \quad a = \frac{R}{2L} \qquad (4\text{-}18)$$

$$\frac{1}{LC} = \omega_n^2 \quad \text{and} \quad \omega_n = \frac{1}{\sqrt{LC}} \qquad (4\text{-}19)$$

For the given resonant system, equation (4-16) becomes

$$i(t) = \frac{E}{\omega_d \cdot L} \cdot e^{-at} \sin \omega_d \cdot t \qquad (4\text{-}20)$$

Pictorially, $f(t)$ is a decaying sine wave, as depicted in Figure 4-9, which describes an oscillating system with a damped frequency $\omega_d = \sqrt{\omega_n^2 - a^2}$ with an exponential rate of decay of e^{-at}, where $a = R/2L$ is referred to as the *damping coefficient* and $\omega_n = 1/\sqrt{LC}$ is the natural resonant frequency.

Numerous other important relationships can be derived in a like manner using the Laplace transformations. Before they are developed, a better understanding of other types of oscillating systems is desirable. These include, in addition to the above, the basic spring–mass system and the closed-loop servo system.

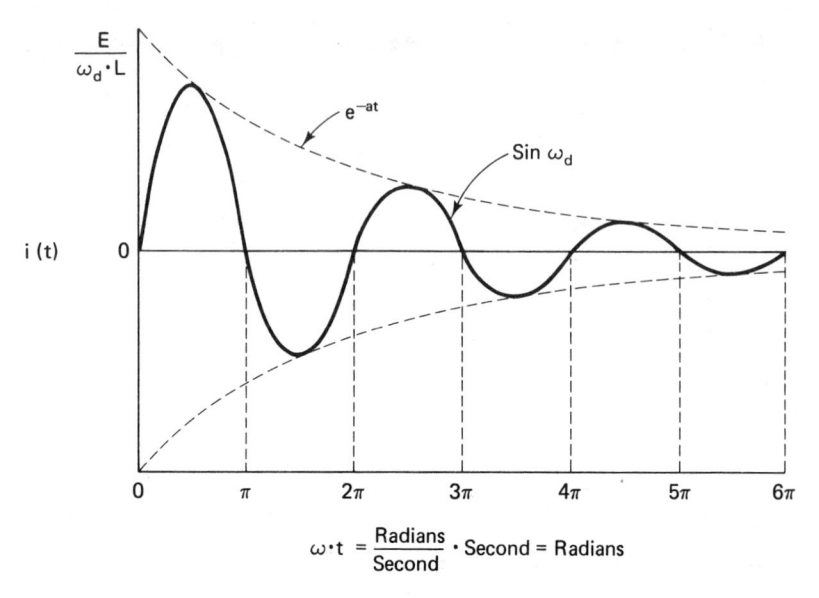

FIGURE 4-9 Damped sine wave $i(t) = E/\omega_d \cdot L \, (e^{-at} \sin \omega_d \cdot t)$.

$\omega \cdot t = \dfrac{\text{Radians}}{\text{Second}} \cdot \text{Second} = \text{Radians}$

The problems at the end of this chapter will provide practice in the use of Laplace transforms from $F(s)$ to $f(t)$. In Chapter 5, actual networks commonly used in automatic control systems will be analyzed using the s operator and the Laplace table.

GLOSSARY

Damped Wave: A series of sine waves decaying at an exponential rate described by e^{-x}, where $x = at$.

Damping Coefficient: The exponential rate of decay of a series of sine waves defined by e^{-at}, where the rate depends on the value of a (the damping coefficient).

Frequency Function: $f(j\omega)$, the output response depends on the steady-state value of the input frequency (ω) to the system. The result is usually in the form $A/\underline{\theta}$.

Function of Time: $f(t)$—the output depends on how long (in real time) the system responds to an input, usually an input of voltage, current, or a mechanical force.

Laplace Transformations: A method of using operational calculus, usually to solve transient problems in which a function of time $f(t)$, becomes, by definition, the Laplace transform $F(s) = \int_0^\infty f(t) \cdot e^{-st} dt$.

s Operator: Basically, the use of s to replace $j\omega$ so that solutions for network characteristics can be accomplished by simple algebra.

Step Input: A function of time where $f(t) = A$, normally 0 for $t < 0$ and A for $t > 0$; in other words, a sudden rise in input at $t = 0$ from 0 to A.

Transient Analysis: The evaluation of the output response of a system to a step-input command signal.

PROBLEMS

Given $f(t)$, determine $F(s)$ using the table of transforms.

4.1 $f(t) = e^{-3t} + e^{-4t}$
4.2 $f(t) = \tfrac{1}{3} e^{-2t}$
4.3 $f(t) = 1 - e^{-t/\tau}$
4.4 $f(t) = \sin 5t$
4.5 $f(t) = 3 \sin 5t$

Given $F(s)$, determine $f(t)$.

4.6 $F(s) = 1/s + 1/(s+3)$
4.7 $F(s) = 1/(3s+6) = A/(s+a)$
4.8 $F(s) = A/s(s\tau + 1)$ for $A = 4$ and $\tau = 0.6$.
4.9 $F(s) = 2/(3s+2)$
4.10 $F(s) = 4/(s^2 + 6s + 100)$ (see pair 9)
4.11 Given TF $= 10/(2s+3)$, determine $A/\underline{\theta}$ at $\omega = 6$ rad/s.

74 The s Operator and the Laplace Transform   Ch. 4

4.12 Given the typical lag network:

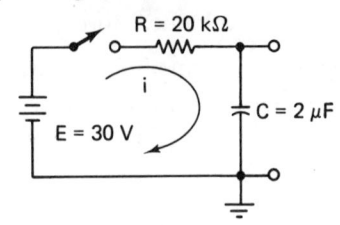

R = 20 kΩ, E = 30 V, C = 2 μF

Determine the current 0.05 s after the switch is closed.

4.13 Given the typical lead network:

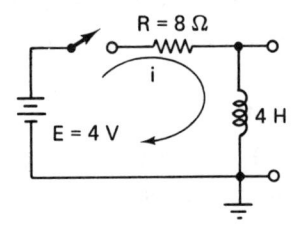

R = 8 Ω, E = 4 V, 4 H

What is the current flow through the coil at $t = 1.5$ s after the switch is closed?

4.14 Given:

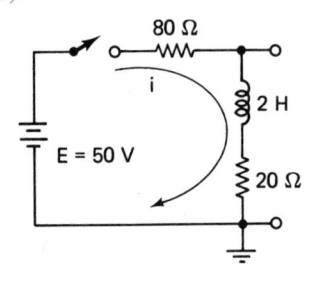

80 Ω, E = 50 V, 2 H, 20 Ω

(a) Solve for $i(t)$ at $t = 0.01$ s. Assume that $t = 0$ at the time of switch closure.

(b) What is the approximate output voltage after five time constants?

4.15 Given the transfer function

$$TF = \frac{20}{s^2 + 6s + 8}$$

What is the output/input response $A/\underline{\theta}$ at $\omega = 5$ rad/s?

4.16 Given

$$TF = \frac{100}{s + 20} = \frac{A}{(s + a)}$$

What is the output as a percentage of the maximum input level 0.2s after the step input is applied?

4.17 From the table of transforms (pair 9), what is the damped frequency (ω_d) when the natural frequency (ω_n) is 100 rad and the damping coefficient (a) is 20?

4.18 Given

$$\text{TF} = \frac{1}{s^2 + 10s + 100}$$

(a) What is the natural frequency (ω_n) of the system?

(b) What is the damped frequency (ω_d)?

(*Hint*: use pair 9 of the Laplace table of transforms.)

5

Solution of Complex Transfer Functions

5-1 INTRODUCTION

In Chapter 4, emphasis was placed on the solution of networks using the s operator and the Laplace Transform pairs. This chapter is an extension of the use of these techniques, with emphasis on the solution of typical automatic control system equations (both open and closed loop) that will be encountered.

Feedback control systems fall into two main catagories, the Type 0 and Type 1 systems. Detailed development of the two basic types is given in Chapters 7 and 8. What is important at this time is that the Type 0 system is a first-order system and can be described by the equation

$$\text{TF} = \frac{b}{s + a} \tag{5-1}$$

whereas the Type 1 system is a second-order system that is described by the equation

$$\text{TF} = \frac{b}{s^2 + 2as + b} \tag{5-2}$$

Since equation (5-2) also describes the behavior of the series resonant *RLC* circuit, the behavior of a spring–mass mechanical resonant system and the behavior of an automatic control system (servo), the importance of understanding the solution of the second-order equation in terms of s cannot be overemphasized.

To complete our understanding of complex networks, specific interpretations of the more common transfer functions in terms of s will be presented.

5-2 SOLUTION OF SECOND-ORDER EQUATIONS IN TERMS OF s

Let us review the solution of a second-order equation in terms of s for $A\underline{/\theta}$ at a frequency ω.

EXAMPLE 5-1

Given the transfer function

$$\text{TF} = \frac{50}{s^2 + 3s + 10}$$

determine $A\underline{/\theta}$ at $\omega = 5$ rad/s.

SOLUTION:

Since an absolute value of ω is given, replacing s by $j\omega$ will result in

$$A\underline{/\theta} = \frac{50}{(j\omega)^2 + 3j\omega + 10} = \frac{50}{j^2\omega^2 + 3j\omega + 10}$$

For $j^2 = -1$ and $\omega = 5$,

$$A\underline{/\theta} = \frac{50}{(-)(5)^2 + 3j(5) + 10} = \frac{50}{-15 + j15}$$

The vector presentation is

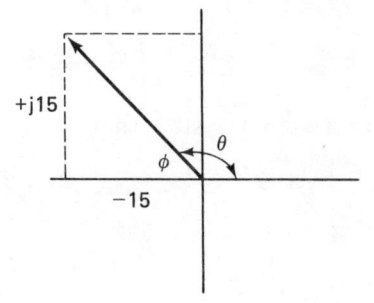

Where $\phi = \tan^{-1}\left(\frac{15}{15}\right) = 45°$

and $\theta = 180° - 45° = +135°$

with the vector $= \sqrt{15^2 + 15^2} = 21.21$

Thus,

$$A\underline{/\theta} = \frac{50}{21.21\underline{/135°}} = 2.36\underline{/-135°} \quad \text{or} \quad +7.45 \text{ dB}\underline{/-135°}$$

The examples that follow are typical transfer functions of an open- and closed-loop system.

EXAMPLE 5-2

The transfer function of an *open-loop servo system* is of the form

$$\text{TF} = \frac{K}{s(s\tau + 1)}$$

With $K = 10$ and $\tau = 0.1$ s, determine $A\underline{/\theta}$ at $\omega = 5$ rad/s. Replacing s by $j\omega$,

$$A\underline{/\theta} = \frac{K}{j\omega(j\omega\tau + 1)} = \frac{K}{j^2\omega^2\tau + j\omega}$$

For $j^2 = -1$, $= K/(-\omega^2\tau + j\omega)$. Substituting $K = 10$, $\tau = 0.1$, and $\omega = 5$,

$$A\underline{/\theta} = \frac{10}{-25(0.1) + j5} = \frac{10}{-2.5 + j5}$$

and

$$A\underline{/\theta} = \frac{10}{5.59\underline{/116.6°}} = 1.79\underline{/-116.6°} = +5 \text{ dB}\underline{/-116.6°}$$

EXAMPLE 5-3

The transfer function of a second-order *closed-loop system* is of the form

$$\text{TF} = \frac{K}{s(s\tau + 1) + K}$$

For the open-loop equation of Example 5-3, what is $A\underline{/\theta}$ at $\omega = 5$ rad/s when the loop is closed?

SOLUTION:

In terms of $j\omega$,

$$\text{TF} = \frac{K}{j\omega(j\omega\tau + 1) + K} = \frac{K}{-\omega^2\tau + j\omega + K}$$

Substituting $K = 10$, $\tau = 0.1$, and $\omega = 5$,

$$A\underline{/\theta} = \frac{10}{-25(0.1) + j5 + 10} = \frac{10}{j5 + 7.5}$$

and

$$A\underline{/\theta} = \frac{10}{9\underline{/33.7°}} = 1.11\underline{/-33.7°} = +0.9 \text{ dB}\underline{/-33.7°}$$

5-3 THE LEAD NETWORK

A group of very important networks used in control system design are of the lead type. As you recall, most devices require time to respond and therefore introduce phase lag in the system. In a closed-loop feedback system, if the phase shift through the system is excessive, the output can be shifted to the point where it is back in phase with the input. As we now know, this shift must not approach 140° (to maintain a phase margin of at least 40°); otherwise, the system can become unstable. To reduce the excessive phase lag ($-\theta$), networks are used to add positive phase shift ($+\theta$) between the input and the output. These networks are referred to as *lead networks*.

Recall the circuit of Figure 5-1(a). In the resistor–capacitor circuit of Figure 5-1(a), the current leads the voltage. Therefore, the output voltage must lead the input voltage by θ since the output voltage is in phase with the current in a resistor. A commonly used lead network is that of Figure 5-1(b), where

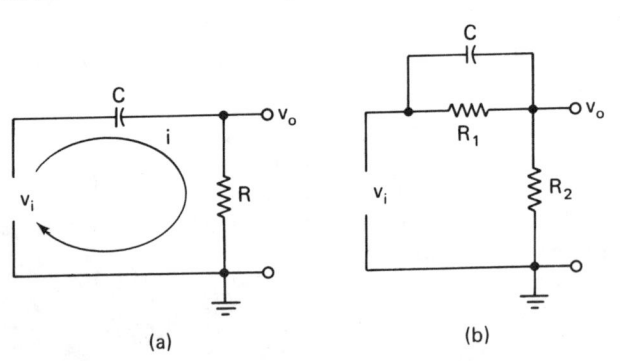

FIGURE 5-1 Two lead networks: (a) basic lead network; (b) popular type lead network used in control systems.

Sec. 5-3 The Lead Network 81

$$\text{TF} = \frac{v_o}{v_i} = \frac{R_2}{R_2 + R_1 \, \mathbb{P} \, 1/sC} \tag{5-3}$$

Mathematically, the solution of equation (5-3) is simplified by solving for $R_1 \, \mathbb{P} \, 1/sC$.

$$R_1 \, \mathbb{P} \, 1/sC = \frac{R_1 \cdot 1/sC}{R_1 + 1/sC} \cdot \frac{sC}{sC} = \frac{R_1}{sCR_1 + 1} \tag{5-4}$$

Thus, the TF of the network of Figure 5-1(b) becomes

$$\text{TF} = \frac{R_2}{R_2 + R_1/(sCR_1 + 1)} \tag{5-5}$$

Multiplying the numerator and the denominator by $(sR_1C + 1)$, equation (5-5) becomes

$$\frac{R_2(sCR_1 + 1)}{sR_1R_2C + R_2 + R_1} \cdot \frac{1/(R_1 + R_2)}{1/(R_1 + R_2)} = \frac{R_2}{R_1 + R_2} \cdot \frac{sR_1C + 1}{\frac{sR_1R_2}{R_1 + R_2}C + 1}$$

and the total transfer function for the lead network is

$$\text{TF} = \frac{R_2}{R_1 + R_2} \frac{(s\tau_1 + 1)}{(s\tau_2 + 1)} \tag{5-6}$$

with $\quad \tau_1 = R_1 C$

$$\tau_2 = \frac{R_1 R_2}{R_1 + R_2} \cdot C$$

This equation shows that $v_o/v_i < 1$ and at low frequencies is equal to $R_2/(R_1 + R_2)$. There are two corner frequencies $1/\tau_1$ and $1/\tau_2$, where $\tau_1 = R_1 C$ and $\tau_2 = [R_1 \cdot R_2/(R_1 + R_2)] \cdot C$. Since τ_2 must be smaller than τ_1, in which case ω_2 is higher than ω_1, the response is then made up of a lead network $(s\tau_1 + 1)$ plus a lag network of $1/(s\tau_2 + 1)$.

EXAMPLE 5-4

Assume that $\tau_1 = 10\tau_2$ with $\omega_1 = 1$ rad/s, then

$$R_1 \cdot C = \frac{10 R_1 R_2}{R_1 + R_2} C \quad \text{or} \quad R_1(R_1 + R_2) = 10 R_1 R_2$$

and
$$R_1 + R_2 = 10R_2 \quad \text{making } R_1 = 9R_2$$

Therefore,
$$K = \frac{R_2}{R_1 + R_2} = \frac{1}{10} = -20 \text{ dB}$$

SOLUTION:

With $\omega_1 = 1/\tau_1 = 1$, then ω_2 must equal $10\omega_1$ or 10 rad/s. The attenuation characteristic of $R_2/(R_1 + R_2) = 1/(1 + 9) = 0.1$, making $K = -20$ dB/$\underline{0°}$.

There are three parts to the Bode plot of the lead network of Figure 5-1(b), which are shown and combined in Figure 5-2. They are:

1. $K = -20$ dB.
2. A lead component of $(s\tau_1 + 1)$, where $\tau_1 = 10\tau_2 = 1$ s.
3. A lag component of $1/(s\tau_2 + 1)$, where $\tau_2 = 0.1\tau_1 = 0.1$ s.

Note in Figure 5-2(d) that the output beyond $1/\tau_2$ is simply a gain of 0 dB/$\underline{0°}$ Also note that by judicious choice of corner frequencies, the phase shift at the *gain crossover frequency* can be modified as much as $+45°$ to help stabilize a design or to improve the phase margin. (See Appendix D for additional details on the use of this particular network for phase compensation.)

Before continuing with network analysis using the s operator, one more point needs mentioning. Since the networks under consideration are *passive devices* made up of resistors, capacitors, and inductors, each will introduce attenuation ($-$dB). Therefore, in most networks, more specifically of the lead type, to bring the plots of A versus θ to a 0-dB reference at very low frequencies requires the use of *active devices* or amplifiers. The added gain ($K = +$dB) depends on the amount of loss ($-$dB) introduced by the network. In Example 5-4, with $\tau_1 = 10\tau_2$, at very low frequencies, where the effect of capacitive reactance is negligible, the output will be down by

$$20 \log_{10} \frac{v_o}{v_i} \quad \text{same as } 20 \log_{10} \frac{R_2}{R + R_2} = -20 \text{ dB}$$

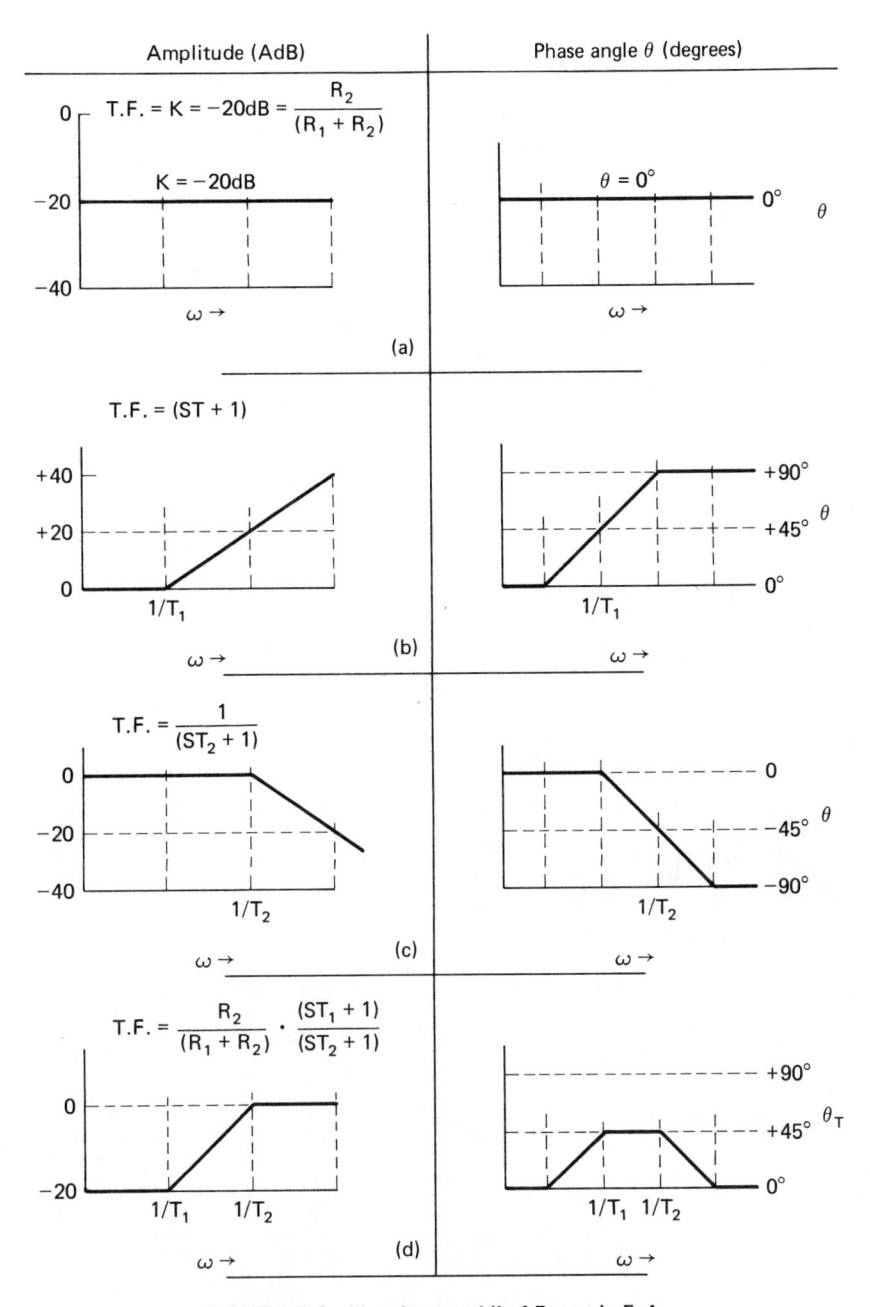

FIGURE 5-2 "Lead network" of Example 5-4.

This loss due to the attenuation characteristics of the network must be overcome by adding amplification to return the gain to unity or 0 dB. The amount of gain must exactly equal the insertion loss, in this case $+20$ dB.

5-4 SPECIAL TRANSFER FUNCTIONS

To complete our understanding of network concepts, several basic functions need interpretation.

1.
$$\text{TF} = j\omega \quad \text{or} \quad s = \omega/\underline{90°} = A/\underline{+90°} \tag{5-7}$$

and for

$$\omega = 1, = 1/\underline{90°} \quad \text{or} \quad 0 \text{ dB}/\underline{90°}$$
$$\omega = 2, = 2/\underline{90°} \quad \text{or} \quad +6 \text{ dB}/\underline{90°}$$
$$\omega = 10, = 10/\underline{90°} \quad \text{or} \quad +20 \text{ dB}/\underline{90°}$$
$$\omega = 100, = 100/\underline{90°} \quad \text{or} \quad +40 \text{ dB}/\underline{90°}$$
$$\omega = 0.1, = 0.1/\underline{90°} \quad \text{or} \quad -20 \text{ dB}/\underline{90°}$$

The Bode plot is simply a straight line with a rising slope of $+20$ dB/decade (at $\omega = 1$, $A = 0$ dB), as shown in Figure 5-3.

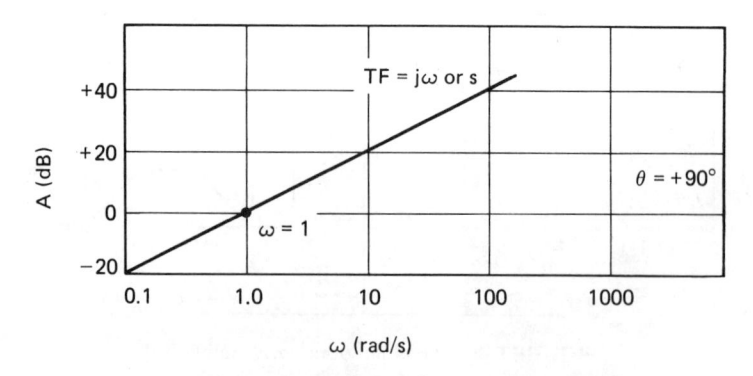

FIGURE 5-3 Bode plot of TF = $j\omega$ or s.

Sec. 5-4   *Special Transfer Functions*   **85**

2.
$$\text{TF} = \frac{1}{j\omega} \quad \text{or} \quad \frac{1}{s} = \frac{1}{\omega}\underline{/-90°} = A\underline{/-90°} \quad (5\text{-}8)$$

and for

$$\omega = 1, = 1\underline{/-90°} = 0 \text{ dB}\underline{/-90°}$$
$$\omega = 2, = 1/2\underline{/-90°} = -6 \text{ dB}\underline{/-90°}$$
$$\omega = 10, = 1/10\underline{/-90°} = -20 \text{ dB}\underline{/-90°}$$
$$\omega = 100, = 1/100\underline{/-90°} = -40 \text{ dB}\underline{/-90°}$$
$$\omega = 0.1, = 10\underline{/-90°} = +20 \text{ dB}\underline{/-90°}$$

The Bode plot is a straight line with a negative slope of -20 dB/decade (at $\omega = 1$, $A = 0$ dB), as shown in Figure 5-4.

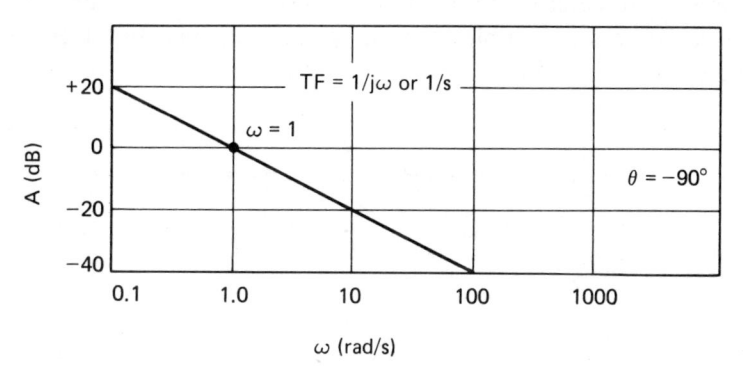

FIGURE 5-4 Bode plot of TF = $1/j\omega$ or $1/s$.

3. The TF = $(j\omega RC + 1) = (j\omega\tau_1 + 1) = (s\tau_1 + 1)$ has a corner frequency of $\omega_c = 1/\tau_1$. At the frequencies beyond which $j\omega RC = 1$ or $\omega_c = 1/RC$, as ω increases, v_o increases proportionately beyond the corner frequency. The output is exactly the reciprocal of $1/(s\tau + 1)$, which we know quite well. In Bode plot form, it appears as shown in Figure 5-5.

When looking at the Bode plots of Figures 5-3, 5-4, and 5-5, a question should immediately come to mind: If the output response

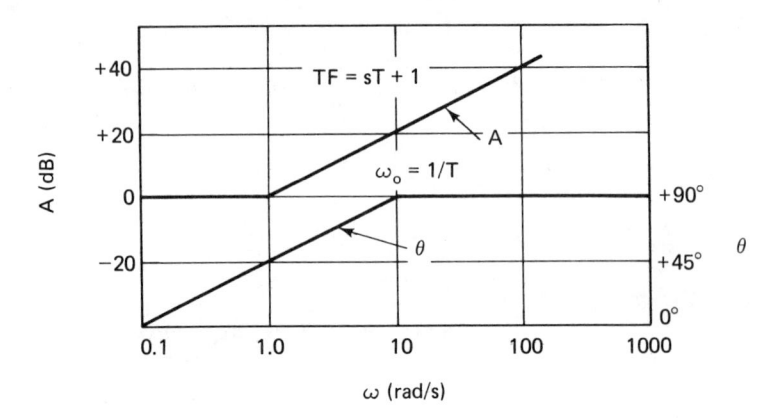

FIGURE 5-5 Bode plot of TF = $(sT + 1)$.

is a +dB, does the system contain amplification? The answer is "yes."

In our study of special functions, particularly s and $1/s$, it is interesting to determine how these responses are actually obtained. Figure 5-6 is an example of an amplifier with negative feedback and a summing point.

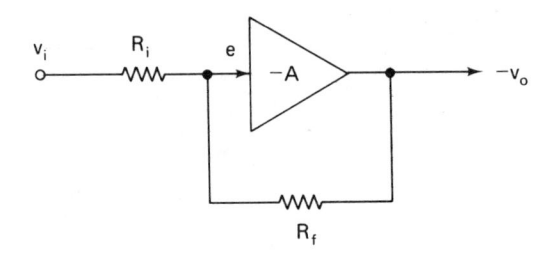

FIGURE 5-6 Basic operational amplifier.

In Chapter 2, it was shown that the transfer function for the operational amplifier using the inverting input is

$$\text{TF} = \frac{-R_f}{R_i} \tag{5-9}$$

A number of operational amplifiers can be combined to make up an analog computer, a device used to simulate the mathematical model of a system to determine the dynamic behavior of the system (without the need of constructing the actual system). By varying the form of feedback, any of many mathematical operations can be performed. These include addition, subtraction, multiplication, division, and the form of calculus known as *differentiation, which in our applications appears as multiplication by s* [see equation (5-7)] or integration, which is multiplying by 1/s [see equation (5-8)]. By replacing R_f in Figure 5-6 with a capacitor as shown in Figure 5-7 (R_f is usually retained to hold the gain to some level determined by the saturation characteristics of the system), it should become apparent that the closed-loop gain is now also a function of frequency, since at zero frequency, $X_C = \infty$ and G is R_f/R_i. As the frequency increases, at $\omega = \infty$, $X_C = 0$ and the gain approaches zero.

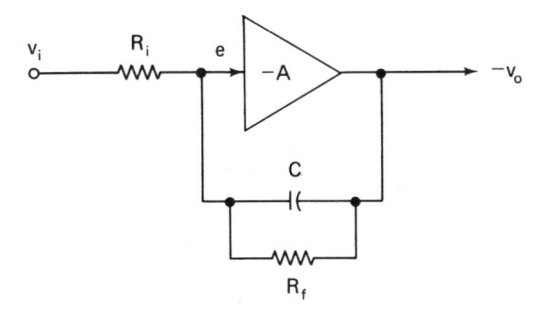

FIGURE 5-7 Integrating amplifier, TF = 1/s.

Mathematically, the transfer function of Figure 5-7, where R_f is replaced by a reactance of $1/j\omega C$, becomes

$$\text{TF} = \frac{1/j\omega C}{R_i} \cdot \frac{j\omega C}{j\omega C} = \frac{1}{j\omega RC} = \frac{1}{s} \cdot \frac{1}{\tau} \quad (5\text{-}10)$$

Thus, we obtain the integration $1/s$ and the gain crossover frequency occurs at a frequency of $1/\tau$ rad/s. In a similar manner, the differentiating amplifier of Figure 5-8 looks similar, except that

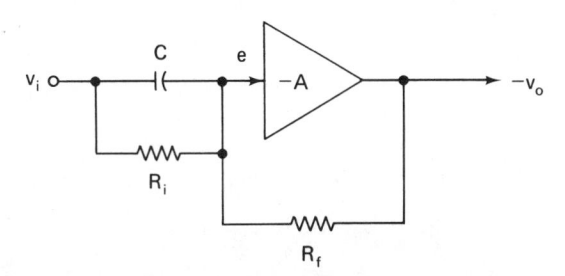

FIGURE 5-8 Differentiating amplifier, TF = s.

the capacitor is placed at the input. In this case, R_i of Figure 5-6 is replaced by $1/j\omega C$ and the transfer function becomes

$$\text{TF} = \frac{R_f}{1/j\omega C} = j\omega RC = s(\tau) \qquad (5\text{-}11)$$

which provides s, the differentiating characteristics described by the plot of Figure 5-3.

The application of operational amplifiers to network or system solutions is quite straightforward, as will be demonstrated.

EXAMPLE 5-5

Take the transfer function given in Example 5-2:

$$\text{TF} = \frac{K}{s(s\tau + 1)}$$

where $\quad K = 10$

$\qquad \tau = 0.1$ s

Determine A/θ at $\omega = 5$ rad/s.

SOLUTION:

The transfer function is made up of three parts:

$$TF = K \cdot \frac{1}{s} \cdot \frac{1}{s\tau + 1}$$

which can be simulated by three operational amplifiers and a series RC lag network with its isolation amplifier to prevent loading of the RC network. The analog computer model is shown in Figure 5-9, where

$\quad K =$ operational amplifier with the feedback R_f adjusted for a gain of $10\times$ so that $R_f/R_i = 10$ or $+20$ dB.

$\quad \dfrac{1}{s} =$ integration amplifier of Figure 5-7 to produce TF $= 1/s$, where $\tau = 1$ s.

$\quad \dfrac{1}{s\tau + 1} =$ simple lag network with $\tau = 0.1$ s

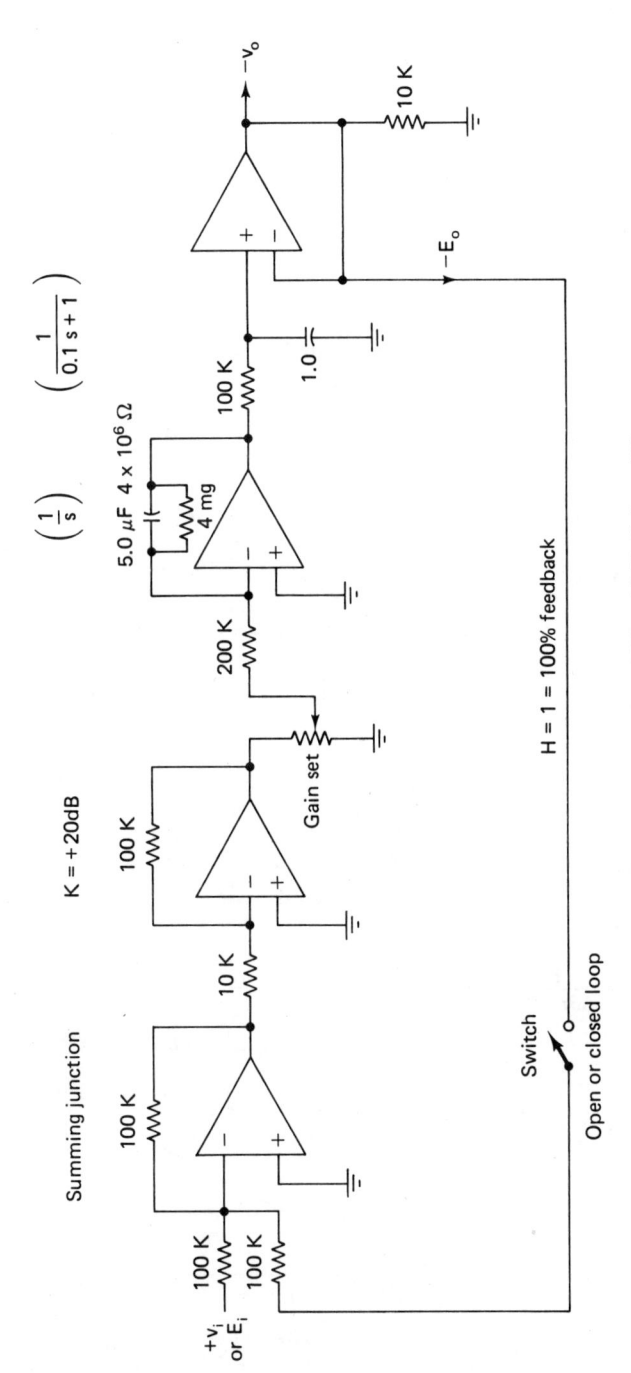

FIGURE 5-9 Analog model of TF = $10/[s(0.1s + 1)]$.

SJ = operational amplifier with a gain of unity having at least two inputs for summing $E_i - E_o$

In practice, the input voltage v_i must be limited to prevent saturation of any of the amplifiers. Also, an odd number of inverting operational amplifiers must be used; otherwise, the output will be back in phase with the input. A look at v_o compared with v_i on an oscilloscope (with v_i at 5 rad/s) would show that the output/input voltage is $+5\text{dB}/\underline{-116.6°}$, as calculated in Example 5-2.

A summary of the basic network responses are shown in Figure 5-10 and should be immediately recognized.

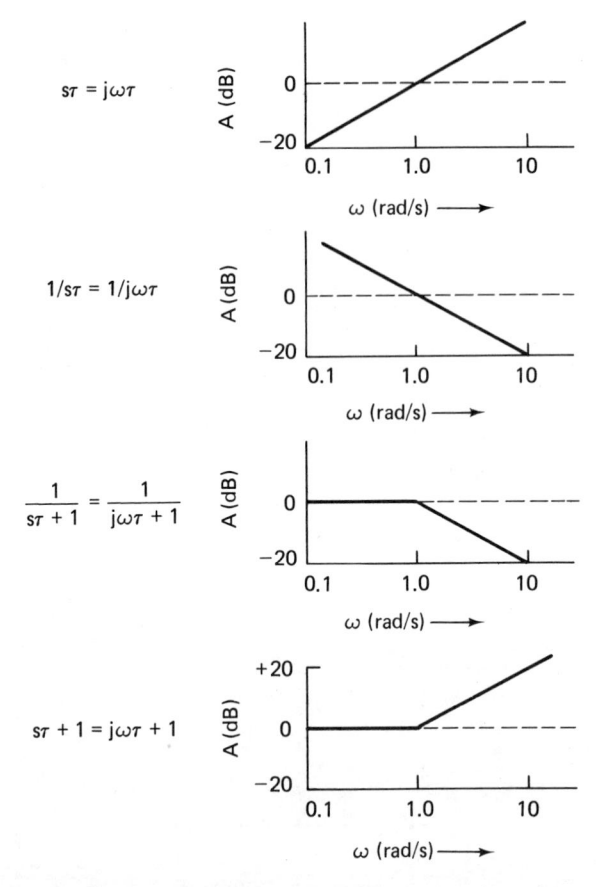

FIGURE 5-10 Basic Bode plots of transfer functions showing amplitude characteristics.

GLOSSARY

Active Network: One using an amplifier with passive components (resistors, capacitors, and/or inductors) to develop a desired frequency-response characteristic.

Analog Computer: A series of "operational amplifiers", each with its feedback adjusted to provide a specific response to simulate the transfer function of a dynamic system.

Integrating Amplifier: An operational amplifier with a resistor R_i as the input using a capacitor C as the feedback network to provide a transfer function of $1/s$.

Lead Network: A group of components arranged to provide an output voltage that has a positive phase angle with respect to the input voltage (taken as the reference). In automatic control systems, it is used to cancel some of the lag or delay introduced because of the usual slow response time of mechanical devices.

Second-Order System: One that by its nature can produce oscillations because it can store energy in two forms. In an electrical system these are the inductance with its magnetic flux and the capacitor with its electrical charge; in a mechanical system, it is the spring under tension and the mass acting against gravity.

Type 1 System: A second-order system; a system also described by the number of integrations $(1/s)$ that appear in the transfer function.

PROBLEMS

In the networks shown, verify the time constants (τ) by solving for the transfer function using the s operator.

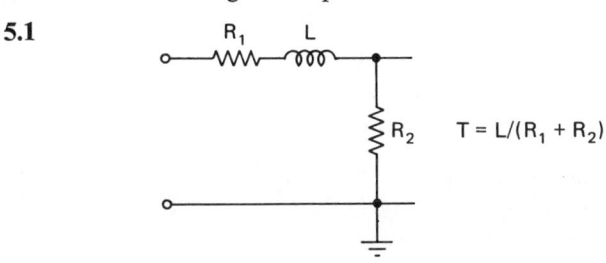

5.1 $T = L/(R_1 + R_2)$

5.2

$T_1 = R_2 \cdot C$
$T_2 = (R_1 + R_2) \cdot C$

5.3

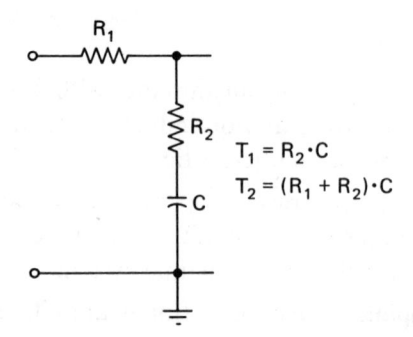

$T_1 = L/R_2$
$T_2 = L/(R_1 + R_2)$

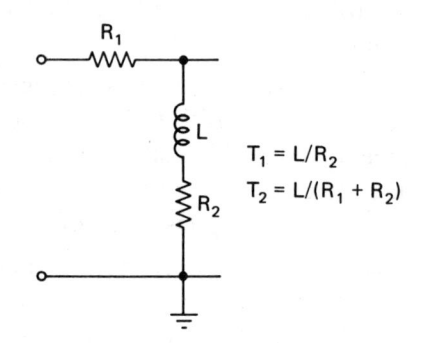

5.4

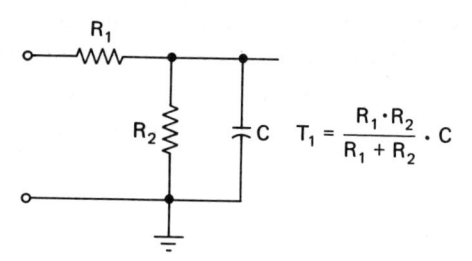

$T_1 = \dfrac{R_1 \cdot R_2}{R_1 + R_2} \cdot C$

5.5 Determine the transfer function in terms of s.

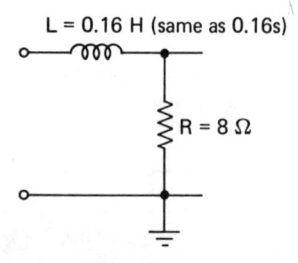

$L = 0.16$ H (same as $0.16s$)

$R = 8\,\Omega$

5.6 What value of C is needed in this circuit to provide the same transfer function as in Problem 5.5?

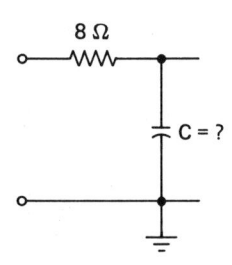

5.7 What are the corner frequencies for the following networks, and are they lead or lag networks?

$$\tau_1 = 0.1 \text{ s}, \tau_2 = 0.5 \text{ s}, \text{ and } \tau_3 = 2.0 \text{ s}.$$

(a) $\dfrac{s\tau_1 + 1}{s\tau_2 + 1}$ (b) $\dfrac{1}{s\tau_3 + 1}$ (c) $s\tau_2 + 1$ (d) $\dfrac{20}{s}$

5.8 Show each of the network responses of Problem 5.7 on a Bode plot.

5.9 Write the transfer function in terms of s and then solve for $A/\underline{\theta}$ at $\omega = 2.5$ rad/s.

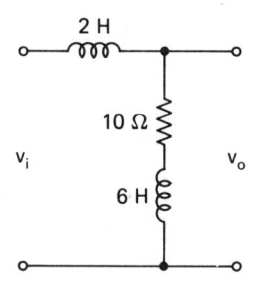

5.10 (a) What is the amplitude and phase of the output of the lead network at $\omega = 10$ rad/s?

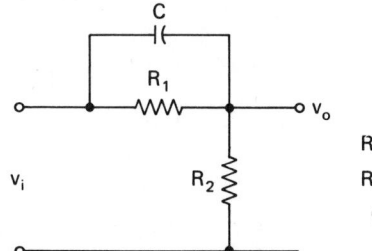

$R_1 = 90 \text{ k}\Omega$
$R_2 = 10 \text{ k}\Omega$
$C = 0.5 \text{ }\mu\text{F}$

(b) Prove your answer (with a Bode plot) of the network response at $\omega = 10$ rad/s.

5.11 Set up an analog computer simulation for the motor transfer function

$$\text{TF} = \frac{10}{s(s\tau_m + 1)}$$

where $\tau_m = 0.4$ s.

6

Control System Components

6-1 INTRODUCTION

Now that we are familiar with the Bode plots, the use of the s operator, and network solutions in terms of $j\omega T$ or sT, we can begin the study of practical systems.

Let us review a typical lag network in combination with an active device as shown in the simple circuit of Figure 6-1, where the lag network has the transfer function $1/(sT+1)$ and the amplifier a gain of K_a. The overall transfer function is the product of the two sections:

$$\text{TF} = \frac{1}{sT+1} K_a \quad \text{or} \quad \frac{K_a}{sT+1} \qquad (6\text{-}1)$$

with $T = R \cdot C$
  $s = j\omega$

Figure 6-1 can be represented by a simple block (part of the flow diagram of a workable control system) as shown in Figure 6-2.

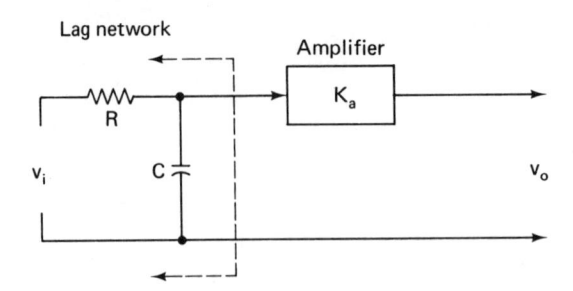

FIGURE 6-1 Transfer function of lag network feeding an amplifier TF = $K_a/(sT + 1)$.

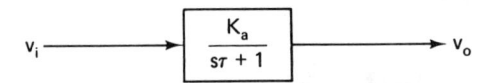

FIGURE 6-2 Block diagram of TF = $K_a/(sT + 1)$.

In a similar way, a series of components can be combined to form a complete control system. Some of the common control system components that will be analyzed are:

Position potentiometer	Servo control valve
Servo motor	Filters
Gear train	Amplifier
Modulator	Position indicators
Demodulator	Tachometer

It is a fact that the number and variety of system components is so numerous that complete texts are available on this subject alone. Special hydraulic, pneumatic, and temperature sensors, which are also important, include:

Strain gauges
Piezoelectric sensors
Stepper motors
Pressure gauges
Solid-state switches
Flow meters
Bimetalic thermostats
Thermocouples

Sec. 6-2 Position Potentiometer 97

and many more. Our goal is to introduce the most basic components for familiarization and establish the transfer function of each. A number of these components will then be incorporated into a system that will be completely analyzed for frequency response and stability. Manufacturers' data will be used, in some cases simplified to fit our needs, to familiarize the student with samples of actual manufactured items.

6-2 POSITION POTENTIOMETER

Control system components are generally divided into two main categories, namely with dc or ac outputs. For example, the position potentiometer of Figure 6-3 is very ruggedly built and is a precision device so as to provide the repeatability and accuracy required of input data. It can be connected in a bridge arrangement such that either dc or ac data are obtainable proportional to shaft position θ.

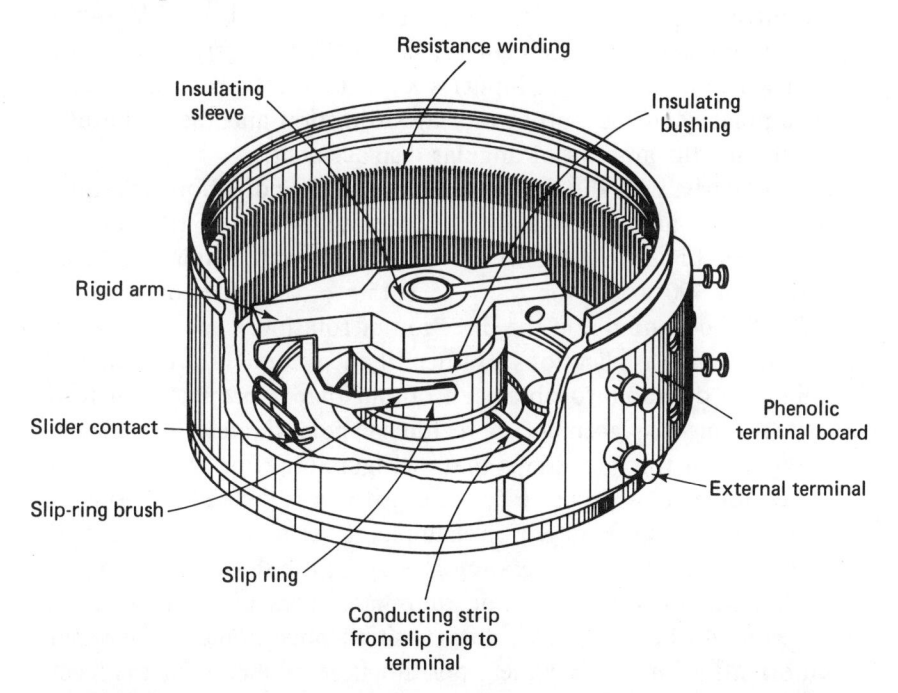

FIGURE 6-3 Position pentiometer. (Courtesy of Bowmar ITIC, Inc.—Subsidiary of Bowmar Instrument Corporation.)

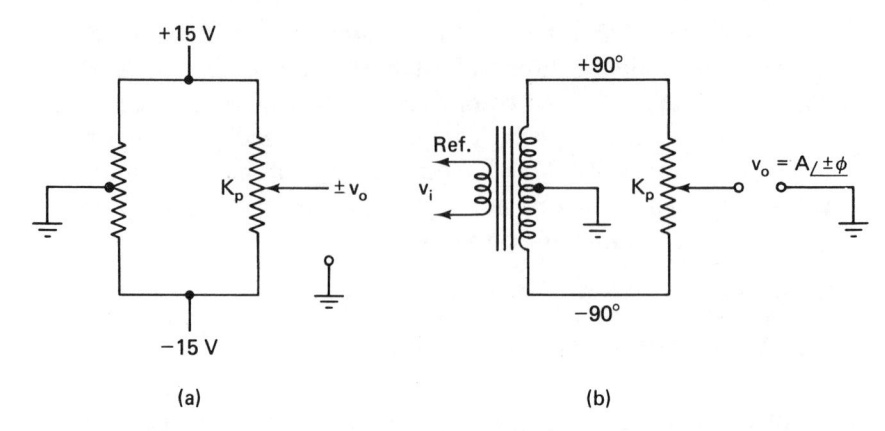

FIGURE 6-4 Dc and ac bridge using a position pentiometer K_p.

In the bridge circuit of Figure 6-4(a), the shaft can be positioned so that at balance, $V_0 = 0$. Depending on the polarity established, a clockwise rotation of the shaft will produce a positive (+) output voltage, and a counterclockwise rotation a negative output. Thus, for the circuit shown, with ±15 V and a maximum shaft displacement of, say, 300°, the sensitivity or the transfer function (output/input) is $K_p = 15 \text{ V}/150° = \pm 0.1$ V/deg. The polarity of the output indicates direction and the amplitude represents the amount of angular displacement.

Similarly, with the total voltage across the potentiometer equal to 30 V ac, at bridge balance the output is zero. Thus, for the circuit of Figure 6-4(b), with a maximum shaft displacement of ±150° or 300° total, the $K_p = 15 \text{ V}/150° = 0.1$ V/deg at an angle of ±90°, depending on the direction of rotation.

In many applications, rather than rotary information, it might be desirable to obtain an output that is proportional to a straight-line displacement. A rectilinear potentiometer specifically designed for this purpose is that of Figure 6-5. For the rectilinear potentiometer of Figure 6-5, the sensitivity is defined in terms of V/in. rather than V/deg. For example, if the total linear displacement is 5 in. and the excitation is ±15 V, the sensitivity or $K_p = 30 \text{ V}/5 \text{ in.} = 6$ V/in. In some automatic control systems, both types might be used: the rotary position potentiometer for input information and the recilinear position potentiometer for the feedback data as used in the system of Figure 6-13.

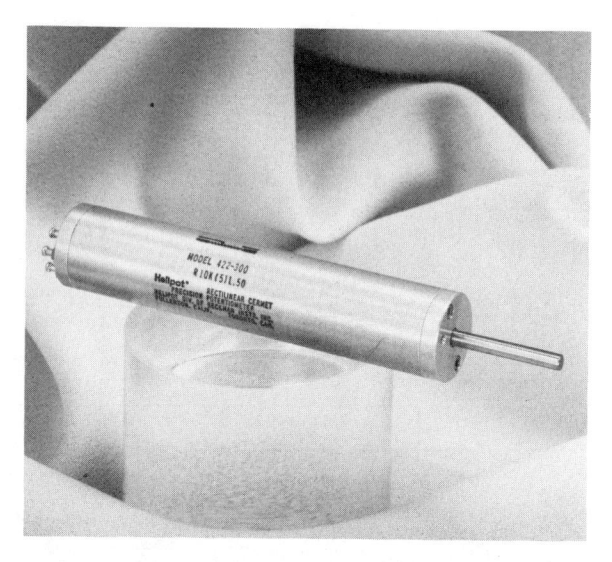

FIGURE 6-5 Rectilinear pentiometer. (Courtesy of Helipot Division of Beckman Instruments, Inc.)

6-3 THE POSITION TRANSFORMER (SYNCHRO)

The study of automatic control system components is not complete without mention of the position transformer, more commonly referred to as the *synchro*. It is basically a device requiring very nearly zero input torque (which is one of its prime advantages) to obtain output voltages that can be transformed into angular position data. It appears in many forms, one of the more common being that of Figure 6-6.

In appearance, the synchro looks like any small motor. The input to the transmitter (also called a *synchro generator*) is a single-phase excitation, usually 110 V 60 Hz. As the rotor is turned, voltage will be induced into the stator windings S_1, S_2, and S_3 (each of which is displaced 120°). The output voltages from the stator of the transmitter are fed to the stator windings of the receiver, which are also placed 120° apart. The magnetic fields created by the stator windings induce a voltage into the rotor of the receiver, depending on its position with respect to each of the three windings.

Depending on the way the receiver is used, the *synchro pair* can

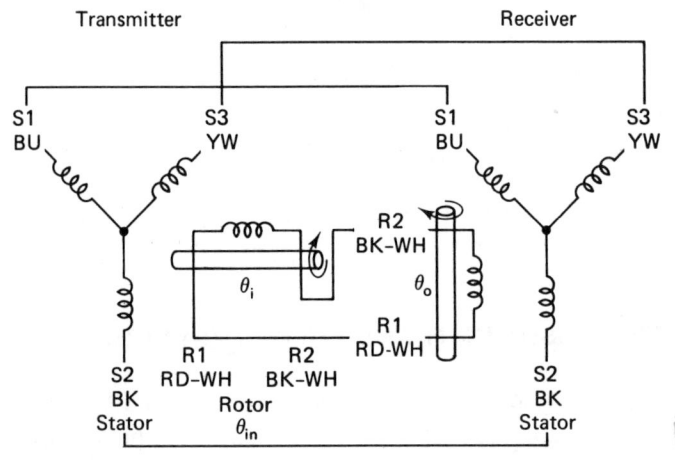

FIGURE 6-6 Synchro "pair". (Courtesy of The Singer Company, Kearfott Division.)

supply an output voltage whose magnitude is proportional to input position with respect to the output position, in which case it is referred to as a *synchro control transformer*. On the other hand, when the rotor of the receiver is excited by the same voltage as that used at the transmitter rotor, the rotor of the receiver will rotate producing a *synchro motor*, in which case the shaft of the receiver will rotate in synchronism with the input and θ_o equals θ_i. The actual torque available at the output is very low and is usually limited to provide position indication only. To obtain high torque requires extensive auxiliary equipment. The output voltage from the receiver is amplified and drives a motor with its own feedback paths; in effect, it becomes an ac type of automatic control system.

6-3-1 The Linear Synchro

However, another form of synchro, called the *linear synchro* is more readily adaptable for the type of application intended in this text. It provides angular data with high resolution with only a slight loading of the driving source (i.e., a compass indication, a gyro position output, or a pen recorder position indication).

The linear synchro is also a small motor-like device similar to Figure 6-6 except that a single "stator" and "rotor" winding are used. The output is simply an ac voltage whose amplitude is proportional to θ_i and is zero when the rotor is at right angles to the stator. Thus, $v_o = K \sin(\theta + 90°)$ or $K \cos \theta$. Simple rectification procedures convert the ac to useful dc "command" information.

6-3-2 Rotary Encoder

The synchros or position transformers described are very simple position indicators compared to the more modern *rotary encoders*. These sophisticated readout devices supply angular information directly in binary numbers. The binary number output is provided by using a series of concentric pattern tracks on a disk similar to Figure 6-7(b). Each track has a photodetector which senses the presence of light or darkness at any given position to provide the binary number, which identifies the absolute position of the disk. In appearance, they look like a small motor with a shaft for supplying input information (θ_i) and an electrical output that can be transmitted to any remote point via wires for direct readout of input angular information. Figure 6-7(a) is an example of a high-resolution rotary encoder used to provide instantaneous electronic information on the position of a rotating shaft or the angular attitude of an instrument such as a camera or radar antenna.

6-4 MODULATOR/DEMODULATOR

The only other special device we will discuss in detail is the modulator demodulator encountered in control system applications. The amplification of low-level dc is a difficult problem because of the dc drift prevalent in most amplifiers. To minimize

102 Control System Components Ch. 6

FIGURE 6-7 Rotary encoder with binary output: (a) typical unit; (b) concentric disk with the binary data. (Courtesy of Teledyne Gurley.)

drift effect and zero unbalance, it is customary to convert the incoming dc to ac (usually by some form of "chopping" of the signal): Figure 6-8(a) is made to look like Figure 6.8(b).

The change from a dc to an ac signal can be done mechanically with a vibrator or electrically by gating a transistor "on" and "off"; the output is then amplified to the desired level. To restore the dc from the ac, a demodulator is used. The demodulator in its simplest form (Figure 6-9) is simply a rectifier made up of a diode and a resistor–capacitor load whose time constant is long enough to provide reasonably good dc output proportional to the peak dc of the amplified modulated input information. The corner frequency due to the *RC* time constant must be well out of range of

Sec. 6-4   Modulator/Demodulator   103

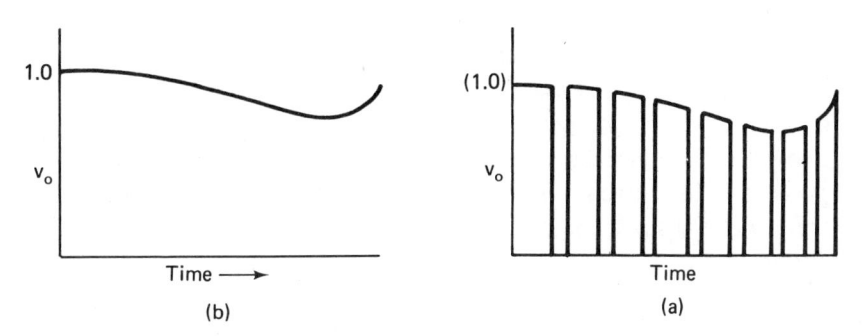

DC$_{in}$ → [Chopper] → [Amp.] → AC$_{out}$

FIGURE 6-8 Modulator: dc to ac (chopped dc).

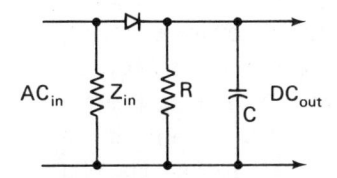

FIGURE 6-9 Simple demodulator circuit.

normal operating frequencies of the system under consideration so as not to add any undesirable phase shift to the system.

The transfer function of the modulator is simply

$$K_m = \frac{\text{ac output}}{\text{dc input}}$$

and for the demodulator,

$$K_D = \frac{\text{dc output}}{\text{ac input}} \tag{6-2}$$

In block form, the overall transfer function for the modulator/demodulator is shown in Figure 6-10, and

$$\text{TF} = \frac{E_o}{E_i} = K_M \frac{V_{ac}}{V_{dc}} \cdot K_D \frac{V_{dc}}{V_{ac}} = K_M \cdot K_D \tag{6-3}$$

E$_i$ → [K$_m$] → [K$_D$] → E$_o$

FIGURE 6-10 Modulator/demodulator where TF = $k_M \cdot k_D$.

6-5 CONTROL SYSTEM DRIVE MOTORS

The final item that is commonly an ac or dc device is the servo motor itself. In its most common form, electrical power is applied to the motor and the output is a rotating shaft whose speed is rpm or rad/s and is dependent on the applied voltage and load.

In many applications, however, the motor appears as a hydraulic valve driving a piston. An input signal opens either of two ports, which causes a linear displacement of the piston proportional to current. This latter method of "positioning" or mechanical displacement is very often used in missile and aircraft applications.

6-5-1 Servo Motor (Angular-Displacement Type)

The servo motor (whose output is a rotating shaft) has either a dc input or is of the usual ac two-phase variety. In many systems, ac is used throughout and the motor is a two-phase induction motor in which a reference voltage is applied to one winding (the reference) and the ac output from an amplifier drives a second winding, called the *control winding*. The phase relationship between the reference voltage and the amplifier output voltage determines the direction of rotation of the motor.

The dc motor appears in two basic forms: either with two field windings, in which case the direction of rotation is governed by the field that has power applied to it, or a conventional motor in which power is applied to the armature and the direction of rotation depends on the polarity of the applied voltage. Figure 6-11 is typical of this type of servo motor.

FIGURE 6-11 Typical dc servo motor. (Courtesy of The Singer Company, Kearfott Division.)

6-5-2 The Servo Control Valve (Linear-Displacement Type)

There are several variations of the servo control valve. Basically, hydraulic fluid under pressure in a storage tank called the *accumulator* is fed into a hydraulic actuator to control rudder position. A positive voltage or current into the valve would result in a "left rudder" movement and a negative voltage or current in a "right rudder" displacement. Figure 6-12 is a good example of a flow control servo valve.

An input current creates magnetic forces on the ends of the armature. A difference in current causes the armature assembly to rotate, moving the flapper assembly, which closes off one nozzle and diverts flow to one end of the spool. This forces the spools to move to one end, opening the pressure side (P_s) to one control port (C_2) and return side (R) to the other control port (C_1), creating a flow to an actuator from C_2 to C_1. The actuator, which is a form

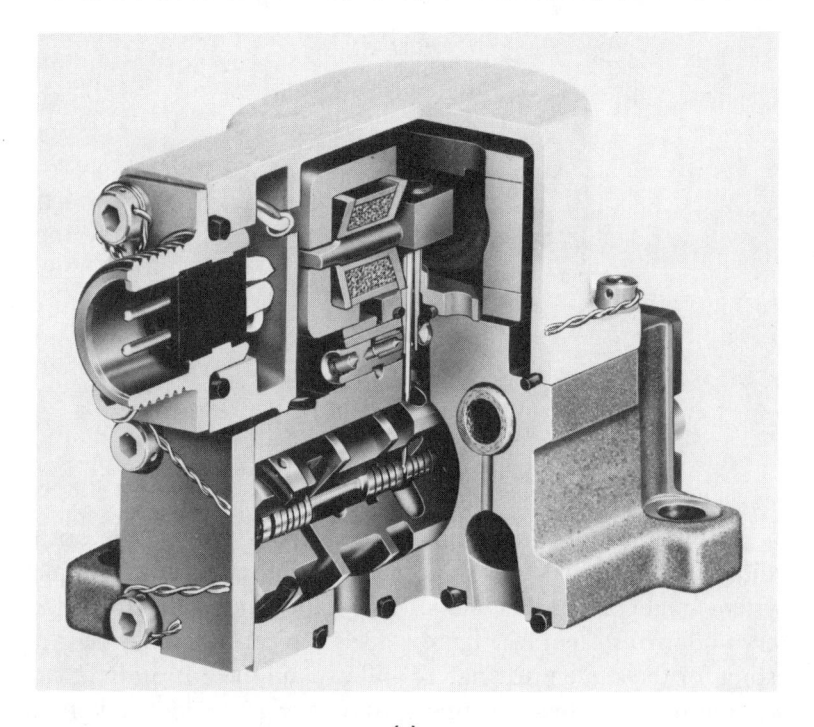

(a)

FIGURE 6-12 Highly sensitive type servo valve. (Courtesy of Moog Inc., East Aurora, N.Y.)

106 Control System Components Ch. 6

(b)

(c)

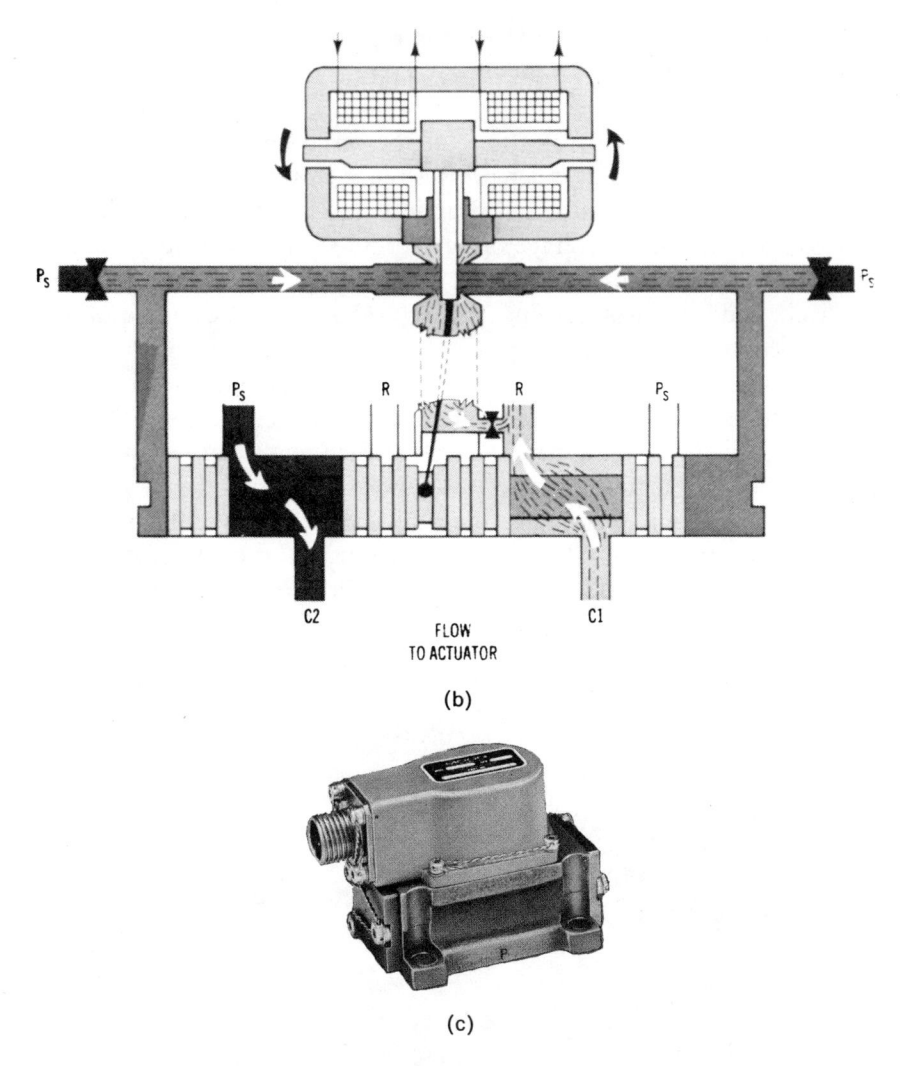

FIGURE 6-12 (Cont.)

of piston, moves a linear potentiometer to provide an output voltage proportional to displacement. In block form, a complete system might appear as shown in Figure 6-13. The closed-loop servo control system of Figure 6-13 is typical. It should be interesting for the student to analyze and explain the operation of each element of the system, starting with a *command signal* and ending with the linear displacement of the *load*.

Since the main objective of the text is to learn the language of

Sec. 6-6 The Transfer Function of a Rotating Type Servo Control **107**

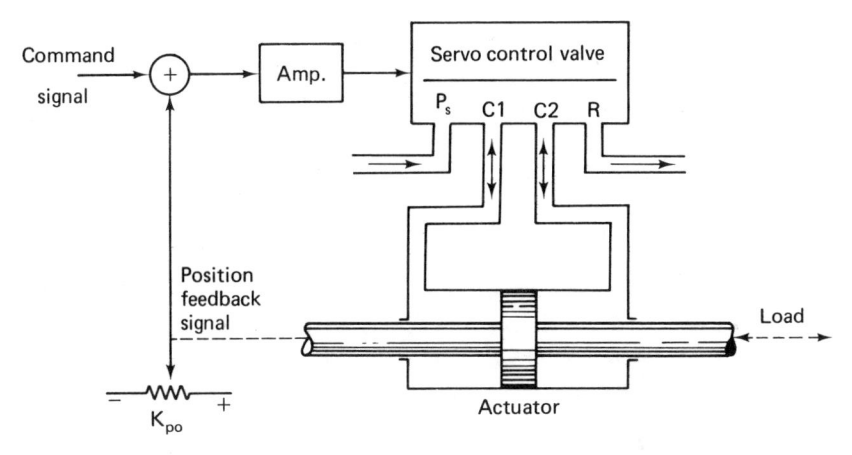

FIGURE 6-13 Closed loop system incorporating a servo control valve. (Courtesy of Moog Inc., East Aurora, N.Y.)

automatic control systems and to be able to perform basic system analysis, the remainder of the text will concentrate solely on the more conventional electrical rotating-type drive motor.

6-6 THE TRANSFER FUNCTION OF A ROTATING TYPE SERVO CONTROL MOTOR

In our system applications, we will concentrate on the dc motor, for a number of reasons:

1. Direct coupling eliminates low-frequency problems associated with capacitor or transformer coupling.

2. Phase compensation is much easier to incorporate by using simple *RC* circuits.

3. Using dc for position and follow-up signals eliminates the need for demodulators and discriminators (phase-sensing device) to establish the error signal (e).

The manufacturer's data for a typical motor such as that of Figure 6-11 is shown in Figure 6-14. Some very important conclusions become evident from the data of Figure 6-14. With 28 V applied to the motor and no load, the motor speed approaches

108 *Control System Components* *Ch. 6*

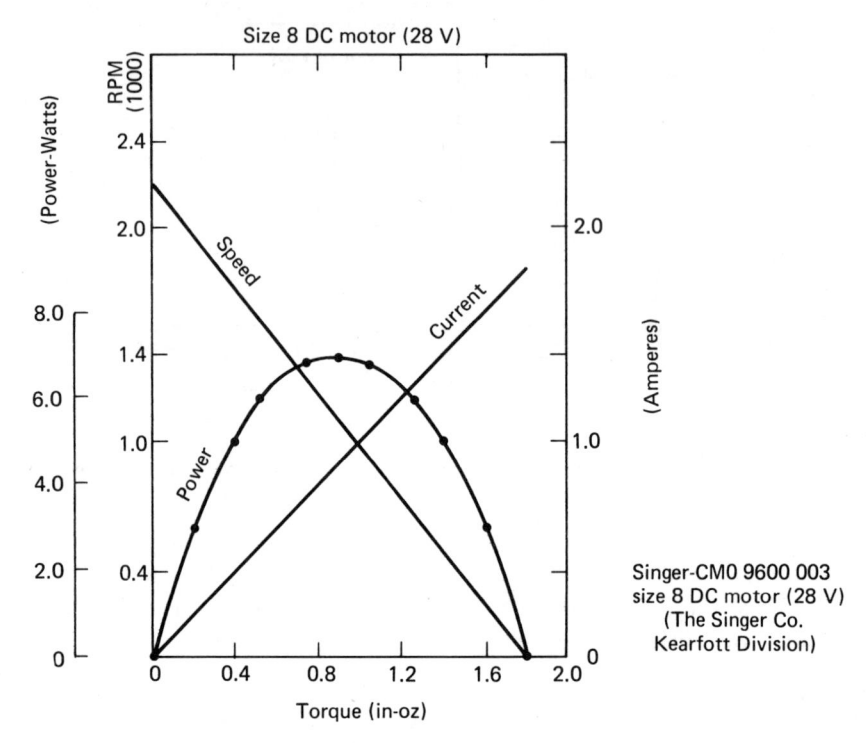

FIGURE 6-14 Manufacturer's data for motor show in Figure 6-11. (Courtesy of The Singer Company, Kearfott Division.)

2200 rpm. Immediately, then, the transfer function of *motor velocity* with rated voltage applied becomes

$$K_{vm} = \text{rpm/V} = \frac{2200}{28} \quad (6\text{-}4)$$
$$= 78.6 \text{ rpm/V}$$

Also from Figure 6-14, with 28 V applied to the motor, the maximum torque is 1.8 in./oz and the transfer function of the output torque versus applied voltage is also quickly established. The *motor torque constant* (K_{tm}) is equal to the maximum output torque divided by the applied voltage:

$$K_{tm} = \frac{1.8 \text{ in.-oz}}{28 \text{ V}} \quad (6\text{-}5)$$
$$= 0.064 \text{ in.-oz/V}$$

Sec. 6-6   The Transfer Function of a Rotating Type Servo Control

We know from Chapter 1 that the servo motor has a basic time constant τ_m. It should be mentioned that other corner frequencies are also present. Since the motor contains an armature or field or both made up of many turns of wire, a motor looks like an inductance and resistance. It should be evident that when power is first applied, the current build up in the field and armature windings are delayed by the ratio of the inductance/resistance (L/R). These electrical time constants are normally many octaves above the basic τ_m of the motor (which was discussed in Section 1-2) and have little affect on the gain crossover frequency or on the phase margin. In our analysis of open- and closed-loop systems, the electrical time constants can be neglected. Thus, only one main corner frequency is present and the complete transfer function of a servo motor is a function of its velocity constant (rpm/volt) or rad/s/V and its time constant τ_m.

$$\text{TF} = \frac{\omega_0}{E_i} = \frac{K_{vm}}{j\omega\tau_m + 1} = \text{rad/s/V} \qquad (6\text{-}6)$$

where ω_0 = rotor speed (rad/s)

E_i = applied voltage

K_{vm} = velocity constant of motor

τ_m = time constant of motor

ω = frequency (rad/s) at which E_i is varied

In equation (6-6), for $\omega = 0$, $\omega_0/E_i = K_{vm}$ = the velocity constant of the motor. Equation (6-6) also includes the frequency characteristic $1/(j\omega\tau_m + 1)$, which contains the corner frequency $1/\tau_m$.

As E_i is varied, the output speed ω_0 also varies. The rate E_i is varied (ω) directly affects the K_{vm}. At $j\omega\tau_m = 1$, the effective K_{vm} is down 0.707 of its steady dc value.

By placing the transfer function in the s domain, the behavior of the motor under varying input is more easily predictable. In the s domain,

$$\omega_0 = s \cdot \theta \text{ (pair 10 of Table 4-1)}$$
$$j\omega = s$$

Equation (6-6) becomes

$$\frac{s \cdot \theta}{E_i} = \frac{K_{vm}}{s\tau_m + 1}$$

and

$$\frac{\theta_o}{E_i} = \frac{K_{vm}}{s(s\tau_m + 1)} \tag{6-7}$$

where θ_o = angular displacement (rad)

E_i = input voltage

The conventional approach for determining the transfer function of a servo motor involves the motor torque constant K_{tm}, the affect on the torque in terms of the moment of *inertia of the motor* (J_m), and the effect of counter EMF as the speed increases on the system damping, referred to as the *viscous damping* (*B*). The complete development of equation (6-7) in terms of *J* and *B* is given in Appendix A.

Some very important observations are presented to recognize the difference in the transfer functions of a motor when it is expressed in terms of its steady-state ($j\omega$) domain compared to the *s* domain. In equation (6-7) we must recognize that E_i is the error voltage (e), whereas in equation (6-6), E_i is a changing dc voltage ($E + V \sin \omega t$). To further clarify the differences between the two equations, ω_o in equation (6-6) is angular velocity, whereas in equation (6-7), θ_o is simply the rotor displacement in degrees or radians making θ_o/E_i the ratio of the rotor displacement with respect to the input voltage.

6-7 TRANSFER FUNCTIONS FOR CONTROL SYSTEM COMPONENTS

The transfer function for the other basic items shown in Table 6-1 are quite apparent. Take the tachometer for example, it is simply a generator whose output is proportional to rpm input, the TF = K_g = V/rpm. In the same way, a gear train such as the type shown in Figure 6-15 can only multiply or divide rpm, and TF is simply the gear ratio *N* or $1/N$.

Sec. 6-7 *Transfer Functions for Control System Components* **111**

TABLE 6-1 Component transfer functions.

Component	Transfer Function
Amplifier	$K_a = V/V$
Demodulator	$K_d = V\ dc/V\ ac$
Gear train	N or $1/N =$ constant
Modulator	$K_m = V\ ac/V\ dc$
Position potentiometer	K_{pi} or $K_{po} = V/\deg$ or V/rad
Servo motor $f(w)$	$K_{vm}/(j\omega\tau_m + 1) = \text{rpm}/V$ or $\text{rad/s}/V$
Servo motor $F(s)$	$K_{vm}/s(s\tau_m + 1) = \theta_o/E_i$
Tachometer $f(w)$	$K_g = E_o/\omega_i = V/\text{rad/s}$ or V/rpm
$F(s)$	$K_g \cdot s = E_o/\theta_i$

FIGURE 6-15 Gear train or speed reducer—TF = 1/N. (Courtesy of Bowmar/ TIC, Inc.—Subsidiary of Bowmar Instrument Corporation.)

The transfer function of items listed in Table 6-1 should be recognizable.

Each of the components or elements that make up a servo system has its own transfer function and must be included when analyzing a complete system. In addition to the gain constant K, each of the transfer functions will also contain a frequency characteristic in terms of $j\omega$ or s. In many of the items listed, there are corner frequencies present which are many octaves beyond the corner frequency of the servo motor itself and therefore have negligible effect at the gain crossover frequency, where the phase margin is determined.

After the transfer function of each block is established and the flow diagram prepared, the system analysis can begin. In the next section, some typical systems will be evaluated using the open-loop approach (determine the Bode plot) to establish the phase margin of the system at the gain crossover frequency. If the phase margin is not within prescribed limits, phase compensation will be added as needed to assure satisfactory performance of the automatic control system when the loop is closed.

6-8 FLOW DIAGRAM OF A SIMPLE CONTROL SYSTEM

Various kinds of networks were set up in the problem section of Chapter 5. These included many types of lead and lag networks and special combinations. We will now show how control system components are combined to make up a complete control system. Figure 6-16 is a flow diagram of a typical servo system such as shown in Figure 6-17, where

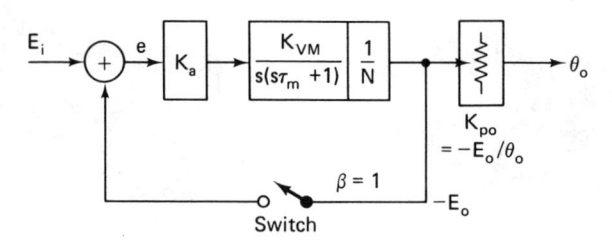

FIGURE 6-16 Flow diagram of a position control system (open-loop).

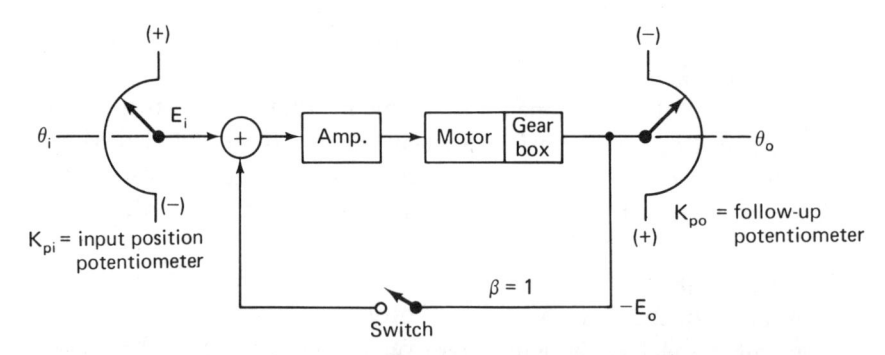

FIGURE 6-17 Typical servo system (open-loop).

K_{pi} = position potentiometer whose output voltage is proportional to shaft position

K_a = wide-band amplifier with a corner frequency too high to be significant, and therefore can be neglected

Sec. 6-9   Phase Compensation   113

$K_{vm}/s(s\tau_m + 1)$ = servo motor characteristics

K_{po} = controlled output device (the follow-up potentiometer)

β = feedback path (the switch position determines whether it is an open or a closed loop)

In addition to the foregoing basic components that make up a closed-loop control system, various types of networks are used for modifying the overall response. For example, if the systems of Figure 6-13 or that of Figure 6-17 were unstable due to too much lag, introduction of lead networks can decrease the overall phase shift to less than (180° − 40°) (when the gain is greater than 1) and the system would be made stable under closed-loop conditions.

6-9 PHASE COMPENSATION

As mentioned in Section 6-8, when the phase margin is not adequate (too much lag is present), circuits that provide leading phase angles are used. Two control elements commonly used specifically for phase compensation and their transfer functions will be included in the list of control system components. They are the lead network, which has been described in Chapters 5 and 6, and the other is the rate generator or tachometer feedback.

The transfer function for Figure 6-18 for $T_1 = 10T_2$ was developed in Chapter 5 and is

$$\text{TF} = \frac{R_2}{R_1 + R_2} \cdot \frac{sT_1 + 1}{sT_2 + 1} \quad \text{with} \quad \frac{R_2}{R_1 + R_2} = -20 \text{ dB} \quad (6\text{-}8)$$

Figure 6-19 shows the linear approximation of the amplitude and phase characteristics of the *lead network* of Figure 6-18. The data show that a lead network of this type can provide approximately +45° phase correction to a system but does it with an insertion loss of 20 dB. Therefore, when a lead network is used for phase compensation, additional gain must usually be added to overcome the insertion loss.

114 *Control System Components* **Ch. 6**

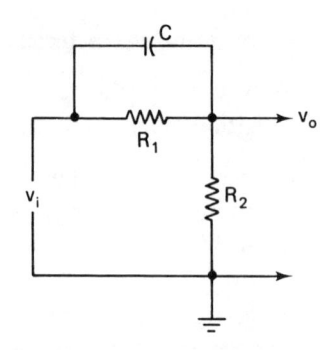

FIGURE 6-18 Lead network
$$\frac{R_2}{R_1 + R_2} = 1/10 = -20 \text{ dB}.$$

$$TF = \frac{R_2}{R_1 + R_2} \cdot \frac{sT_1 + 1}{sT_2 + 1}$$

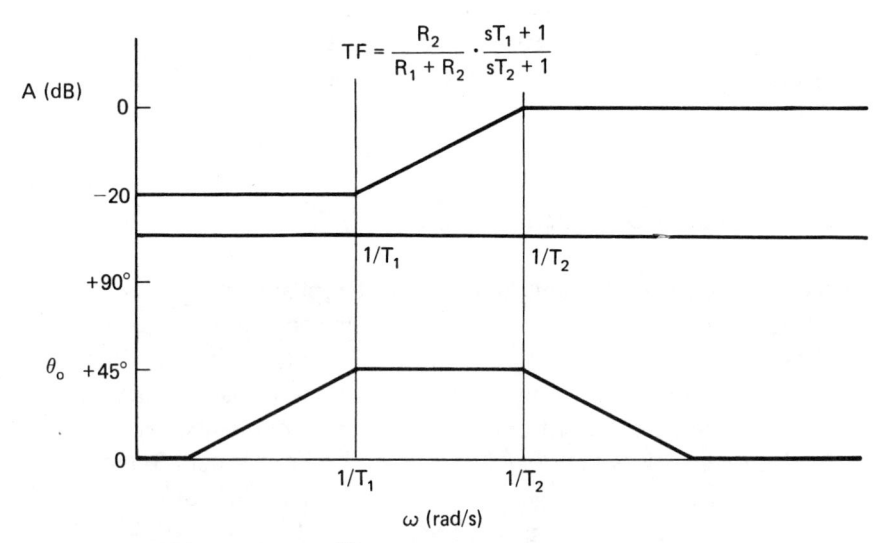

FIGURE 6-19 $A/\underline{\theta}$ plot of "lead network" with $T_1 = 10T_2$.

6-10 THE RATE GENERATOR

Another interesting device for adding positive phase shift is the *rate generator*. This instrument is simply a tachometer whose output is proportional to rpm or rad/s. Figure 6-20 is a typical unit. The transfer function of a tachometer from Table 6-1 is

$$TF = K_g = \frac{v_o}{\omega_i}$$

or

$$v_o = K_g \cdot \omega_i \qquad (6\text{-}9)$$

Sec. 6-10 The Rate Generator 115

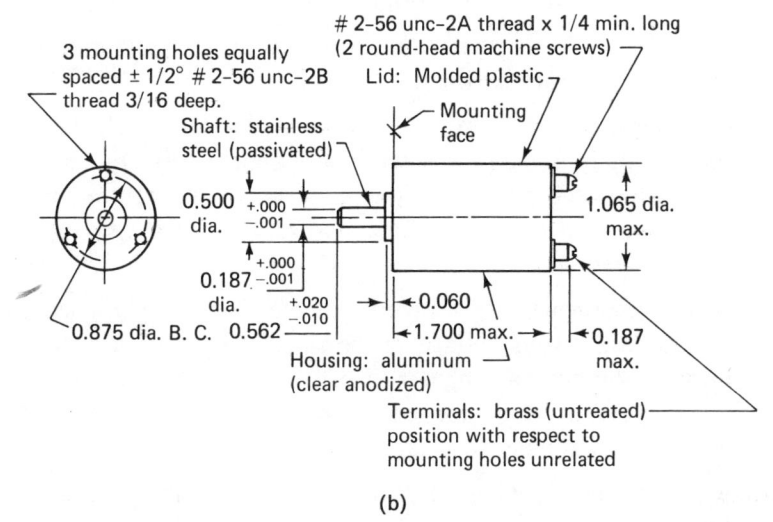

FIGURE 6-20 Tachometer or dc generator—TF = $\frac{V}{rad/s}$ (Courtesy of Helipot Division of Beckman Instruments, Inc.)

In the s domain, ω_i can be replaced by $s \cdot \theta$ (pair 10, Table 4-1) and

$$E_o = K_g \cdot s \cdot \theta_i$$

from which

$$\text{TF} = \frac{E_o}{\theta_i} = K_g \cdot s \qquad (6\text{-}10)$$

A look at the transfer function shows that when s is replaced by

$j\omega$, $K_g \cdot s$ becomes $K_g \cdot j\omega$ or $K_g \cdot \omega\underline{/+90°}$. Thus, by the use of rate feedback compensation, as much as $+90°$ of phase shift can be added to the feedback loop, Mathematically, then, since multiplication by s is differentiation, equation (6-10) indicates that a tachometer is actually a differentiator, when added to a feedback loop, it can nullify the integration $1/s$ present in a motor transfer function.

GLOSSARY

Demodulator: Converts ac (usually the output from a modulator) to dc, retaining the amplitude characteristics of the input.

Modulator: A device that converts dc to ac still retaining the amplitude characteristics.

Phase Compensators: Circuit elements or devices which (in our specific analysis) add a positive phase angle ($+\theta$) to a control system loop to improve the phase margin. Usually, it is in the form of a lead network or is a rate generator used to supply velocity feedback.

Position Potentiometer: A precision potentiometer designed to provide an output voltage proportional to input mechanical information such as angular displacement in degrees. Also, a linear potentiometer for converting linear displacement in inches or centimeter to volts.

Rate Generator: A tachometer used to supply velocity feedback to an automatic control system to improve the phase margin.

Synchro Receiver: Converts electrical information from a synchro transmitter into mechanical date (angular position) or into an ac output voltage whose amplitude and phase depend on the angular position of the input to the transmitter.

Synchro Transmitter: A position transformer with a set of fixed windings and one rotating winding. The output is a set of voltages proportional to the position of the rotated winding. Basically, it converts angular position information into an output voltage with a variable phase angle.

Torque Constant of a Motor: K_{tm}, the torque characteristic of a motor; usually in in.-oz/V or dyn/cm/V.

Velocity Constant of a Motor: K_{vm}, the sensitivity of a motor expressed in rpm/V, or in the metric system in terms of rad/s/V.

PROBLEMS

6.1 Given a potentiometer with 50 V dc across it, what is the transfer function of the potentiometer (K_p) if the total usable rotation is 250°? Give your answer in V/rad.

6.2 The potentiometer of Problem 6.1 is fed into an amplifier with a gain of 12:

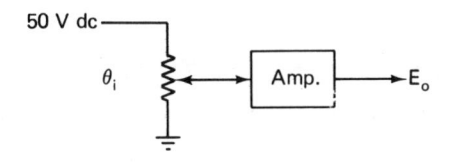

(a) Draw the block diagram of the two elements of Problem 6.2.

(b) What is the transfer function?

(c) What is the output voltage when $\theta_i = 30°$?

6.3 Given the flow diagram:

$$K_x = \frac{2}{sT_x + 1} \quad \text{where } T_x = 0.05 \text{ s}$$

$$K_a = 15$$

$$K_{vm} = \frac{5}{sT_m + 1} \quad \text{where } T_m = 0.2 \text{ s}$$

(a) What is the overall transfer function?

(b) What are the corner frequencies?

118 Control System Components Ch. 6

6.4 Convert 1 rpm/V to rad/s/V.

6.5 Given the following servo motor characteristics:

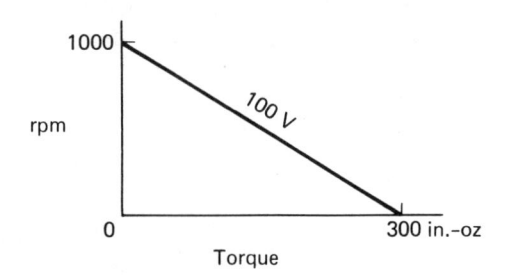

[Graph: rpm vs Torque, with 1000 rpm at 0 torque, line labeled 100 V going to 300 in.-oz at 0 rpm]

    (a) Determine the K_{vm} in rpm/V.

    (b) Determine the torque constant K_{tm} in in.-oz/V.

6.6 Given:

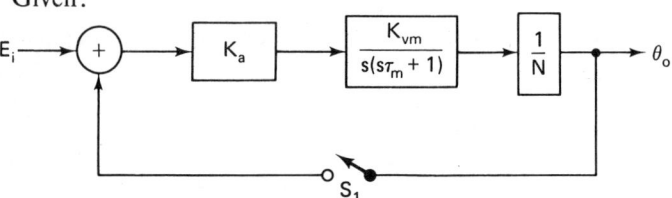

[Block diagram: E_i → summing junction → K_a → $\frac{K_{vm}}{s(s\tau_m + 1)}$ → $\frac{1}{N}$ → θ_o, with feedback through switch S_1]

Write the transfer function of the servo system shown open loop (i.e., θ_o/E_i).

6.7 Given a servo motor with a time constant of 0.1 s and $K_{vm} = 100$ rad/s/V, write the transfer function of the motor in a position control application.

6.8 Given the system:

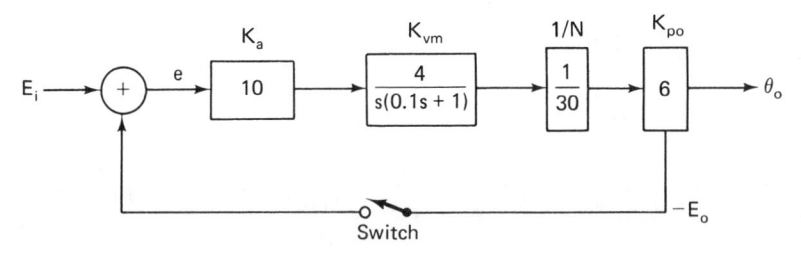

[Block diagram: E_i → summing junction → e → $K_a = 10$ → $K_{vm} = \frac{4}{s(0.1s+1)}$ → $1/N = \frac{1}{30}$ → $K_{po} = 6$ → θ_o, with feedback $-E_o$ through Switch]

    (a) Write the transfer function (E_o/E_i) obtained with the feedback switch open.

    (b) If the motor has a second time constant due to the armature winding (L/R of 0.01 s), write the open-loop transfer function for a position control application including the electrical time constant of the motor, K_{vm}, K_a, the gear box, and the follow-up potentiometer.

6.9 Given the lead network:

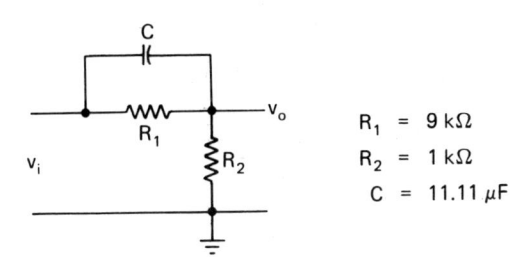

$R_1 = 9 \text{ k}\Omega$
$R_2 = 1 \text{ k}\Omega$
$C = 11.11 \text{ μF}$

(a) What is the overall transfer function?
(b) What is the insertion loss in decibels?
(c) What are the corner frequencies in rad/s?
(d) Show the amplitude and phase plot using semilog graph paper.

6.10 If a tachometer is to be used as a rate generator and has an output of 10 V/1000 rpm input, what is the transfer function of this tachometer as a rate generator?

6.11 Referring to the 28-V, 2200-rpm motor characteristics shown in Figure 6-14, if power output equals rpm × torque, at what speed is the power output maximum with rated input voltage?

7

Open-Loop System Analysis

7-1 INTRODUCTION

In Chapter 3, the stability of a *closed-loop amplifier* was analyzed using its *open-loop equation* and the Bode plot to find the gain crossover frequency and the phase margin.

An automatic control system is a closed-loop system in which the output is fed back to the input with the objective of reducing the error signal to zero. A simple position control system in terms of its output/input (θ_o/θ_i) can be depicted by the closed-loop system of Figure 7-1.

In many ac systems, a synchro transmitter supplies the input signal, which is proportional to the input position θ_i. The output drives another synchro, whose output is proportional to θ_o. The error signal (e) is derived from a phase detector to provide a voltage directly proportional to ($\theta_i - \theta_o$).

The error voltage (e) is amplified and drives the motor, which drives the output synchro until the error voltage from the phase detector is not sufficient to drive the motor.

122 *Open-Loop System Analysis* Ch. 7

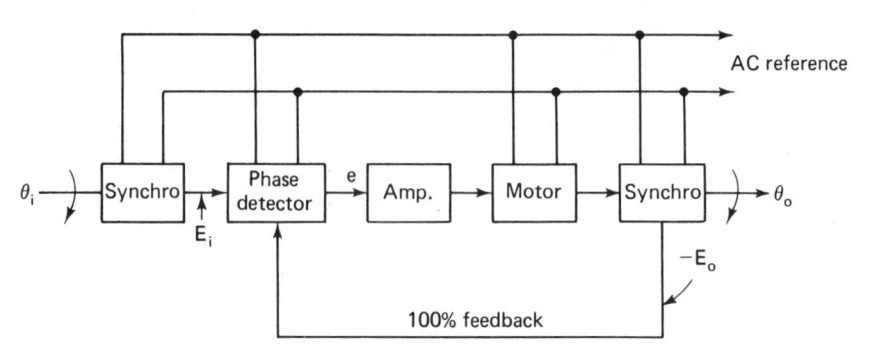

FIGURE 7-1 Typical ac-type position control system.

The forward gain of the system (K) can be defined in terms of the error signal as

$$K = \frac{E_o}{E_i - E_o} \quad (7\text{-}1)$$

or in terms of angular displacement as

$$K = \frac{\theta_o}{\theta_i - \theta_o} \quad (7\text{-}2)$$

and for $e = \theta_i - \theta_o$,

$$K = \frac{\theta_o}{e} \quad (7\text{-}3)$$

Adding the frequency characteristics (G) of the system, the open-loop transfer function becomes

$$\frac{\theta_o}{e} = K \cdot G$$

where $K =$ product of the forward gain of each element in the system

$G =$ frequency characteristic of the system which consists primarily of the motor, $1/[s(s\tau_m + 1)]$

and the overall open-loop transfer function* is

$$\text{TF}_{\text{OL}} = \frac{\theta_o}{e} = \frac{K}{s(s\tau_m + 1)} \quad (7\text{-}4)$$

* Appendix A produces this most important equation using moment of inertia and viscous damping to establish the transfer function.

7-2 POSITION CONTROL SYSTEM TRANSFER FUNCTION

The ac system shown in Figure 7-1 is much more complicated than an all-dc system because of the need to maintain an ac reference voltage in all the subsystems.

A dc control system performs as well as an ac system but is much simpler to analyze. Therefore, the all-dc system will be used as the working model for most of the open- and closed-loop system analysis that follows. A basic dc control system is that of Figure 7-2. A potentiometer (K_{pi}) with dc excitation provides a

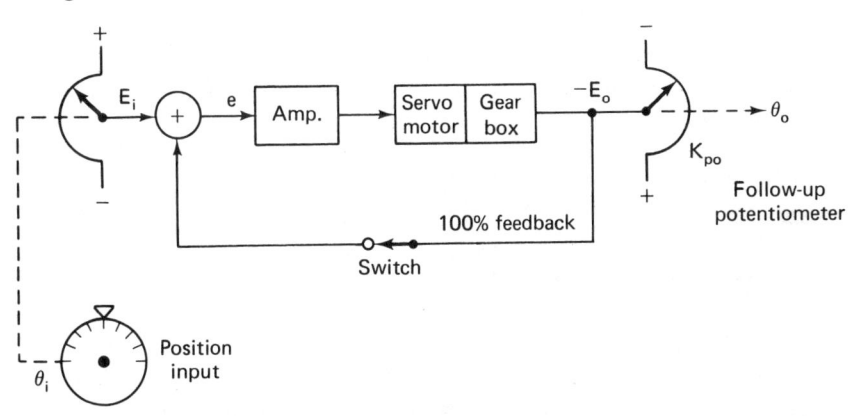

FIGURE 7-2 DC Position control system.

voltage E_i proportional to θ_i, and an output potentiometer (K_{po}) connected to the output shaft of the motor or to the output of a gear box provides E_o proportional to θ_o.

A dial marked in degrees is rotated to provide a position input θ_i (also E_i) as the command signal. The input is fed to an amplifier that drives a servo motor. The output of the motor drives a follow-up potentiometer K_{po}, the output of which goes negative ($-E_o$) for a positive ($+E_i$) input signal. With negative feedback, as ($E_i - E_o$) approaches zero, the motor stops since the input to the amplifier approaches zero and there is no longer drive power to the motor.

The gain (K) of the system of Figure 7-2, like that of Figure 7-1, is equal to the output voltage E_o divided by the error voltage e, so that

$$K = \frac{E_o}{e} = \frac{E_o}{E_i - E_o} \quad \text{[the same as (7-2)]}$$

where $K =$ forward gain of the system
$$E_i = K_{pi} \cdot \theta_i$$
$$E_o = K_{po} \cdot \theta_o$$

Substituting $K_{pi} \cdot \theta_i$ for E_i and $K_{po} \cdot \theta_o$ for E_o,

$$K = \frac{K_{po} \cdot \theta_o}{K_{pi} \cdot \theta_i - K_{po} \cdot \theta_o} \tag{7-5}$$

With $K_{pi} = K_{po}$

$K = \theta_o/(\theta_i - \theta_o)$ [the same as equation (7-3)]

Substituting e for $(\theta_i - \theta_o)$ and adding the frequency characteristic of the motor, the open-loop transfer function for the dc system of Figure 7-2 is also

$$\text{TF}_{\text{OL}} = \frac{\theta_o}{e} = \frac{K}{s(s\tau_m + 1)} \tag{7-6}$$

7-2-1 Bode Plot Analysis

Like any amplifying system with negative feedback, when the frequency characteristic G is added, the phase shift through the system must not exceed 140° at the 0-dB gain point in order to provide a desirable phase margin of at least 40°. If the phase margin is not within limits, phase-compensation techniques must be introduced into the system to improve the phase margin.

With this in mind, open-loop analysis of several systems using the Bode plot will be made and the phase margin determined. The form of phase compensation will be limited to the use of the two common techniques mentioned in Chapter 4:

1. Lead network.
2. Rate feedback.

Before beginning system analysis, it must be emphasized that the standard units of measurement are in the CGS system:

Radians

Dynes

Centimeters

Grams

Seconds

Ohms

Henries

Farads

Volts

Amperes

Several examples will demonstrate the form transfer functions must be presented to fit into the equations that have been developed.

EXAMPLE 7-1

A motor characteristic is given as 2000 rpm with a rated input of 10 V, what is its speed characteristic (K_{vm})?

SOLUTION:

$$K_{vm} = 2000 \text{ rpm}/10 \text{ V} = 200 \text{ rpm/V}$$

To convert 200 rpm/V to rad/s/V = (200 rpm) (2π rad/60 s) and

$$K_{vm} = \frac{200 \times 6.28}{60} = 20.94 \text{ rad/s/V}$$

EXAMPLE 7-2

A potentiometer with a total travel of 150° is energized with 50 V. Determine the transfer function of the potentiometer in V/rad.

SOLUTION:

$$K_p = 50 \text{ V}/150° = 0.3 \text{ V/deg}$$

Since 1 rad = 57.3°, in V/rad,

$$K_p = 0.3 \text{ V/deg} \times 57.3 = 17.3 \text{ V/rad}$$

EXAMPLE 7-3

The torque constant (K_t) of a servo motor is given in in.-oz/V. Convert the torque to dyn-cm/V.

126 *Open-Loop System Analysis* *Ch. 7*

SOLUTION:

$$1 \text{ in.} = 2.54 \text{ cm}$$
$$1 \text{ oz (force)} = 2.78 \times 10^4 \text{ dyn}$$

Therefore,

$$K_t = (2.54 \times \text{in.})(2.78 \times 10^4 \text{ oz-V})$$
$$= 7.06 \times 10^4 \text{ dyn-cm/V}$$

Figure 7-3 is a block diagram of the system of Figure 7-2 with the control system components inserted into the system blocks.

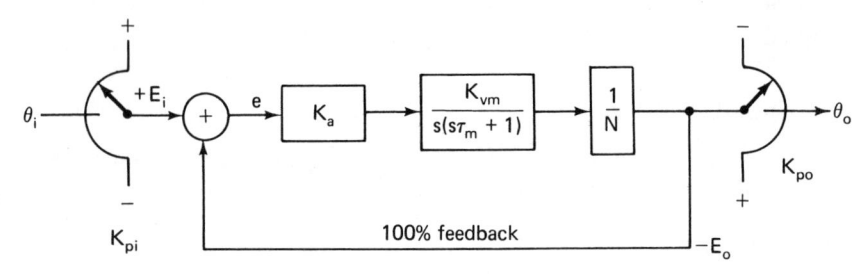

FIGURE 7-3 Block diagram of a dc position control system.

For purpose of analysis, let us give the blocks of Figure 7-3 a set of values:

$$K_a = 10 \text{ V/V}$$
$$K_{vm} = 20/[s(0.1s + 1)] \text{ rad/s/V}$$
$$\beta = 100\% \text{ feedback} = 1$$
$$K_{pi} = K_{po} = 5 \text{ V/rad}$$
$$N = 10$$

Figure 7-3 with the component transfer functions inserted becomes Figure 7-4.

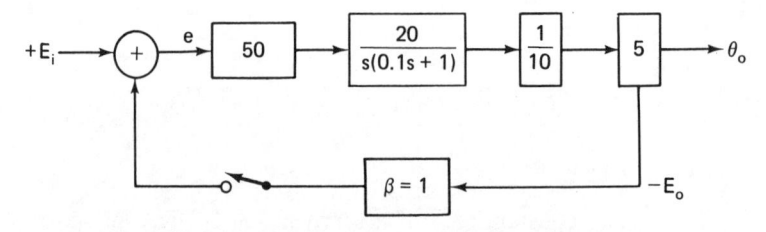

FIGURE 7-4 System of Figure 7-3 with component transfer functions inserted.

Sec. 7-2 Position Control System Transfer Function 127

The open-loop transfer function is readily determined to be

$$\mathrm{TF_{OL}} = \frac{K_a \cdot K_{vm} \cdot K_{po}/N}{s(s\tau_m + 1)}$$

$$= \frac{K_o}{s(s\tau_m + 1)} \quad (7\text{-}7)$$

where K_o = product of the overall forward loop gain = $K_a \cdot K_{vm} \cdot K_{po}/N$. Substituting values into equation (7-7),

$$\mathrm{TF} = \frac{10 \times 20 \times 5/10}{s(s\tau_m + 1)} \quad (7\text{-}8)$$

and the open-loop transfer function for the system of Figure 7-4 is

$$\mathrm{TF} = \frac{100}{s(0.1s + 1)} \quad (7\text{-}9)$$

Our first concern is to determine the phase margin ϕ and the degree of compensation needed to assure satisfactory performance *before* the loop is closed. Figure 7-5 shows the three parts of equation (7-9) that make up the system of Figure 7-4.

The transfer function of equation (7-9) is a typical open-loop transfer function of a basic servo system which includes the overall gain K_o and the frequency characteristic $1/s$ and $1/(s\tau_m + 1)$ in combination. It is well to point out there is an alternative approach to that of plotting the K_o as a decibel gain at $\omega = 1$, as done in Figure 7-5. It may be more convenient to combine $K_o \cdot 1/s$ as K_o/s times $1(s\tau_m + 1)$. For example, instead of treating $1/s$ as

$$\frac{1}{s} = \frac{1}{j\omega} = \frac{1}{\omega}\underline{/-90°} \quad (7\text{-}10)$$

it might be more convenient to combine the gain (K_o) with $1/s$. In the case of the transfer function of equation (7-9), it can be broken into two parts:

$$\mathrm{TF} = \frac{100}{s} \cdot \frac{1}{0.1s + 1} \quad (7\text{-}11)$$

where $100/s$ in terms of $j\omega$ is

$$\frac{100}{j\omega} = \frac{100}{\omega}\underline{/-90°}$$

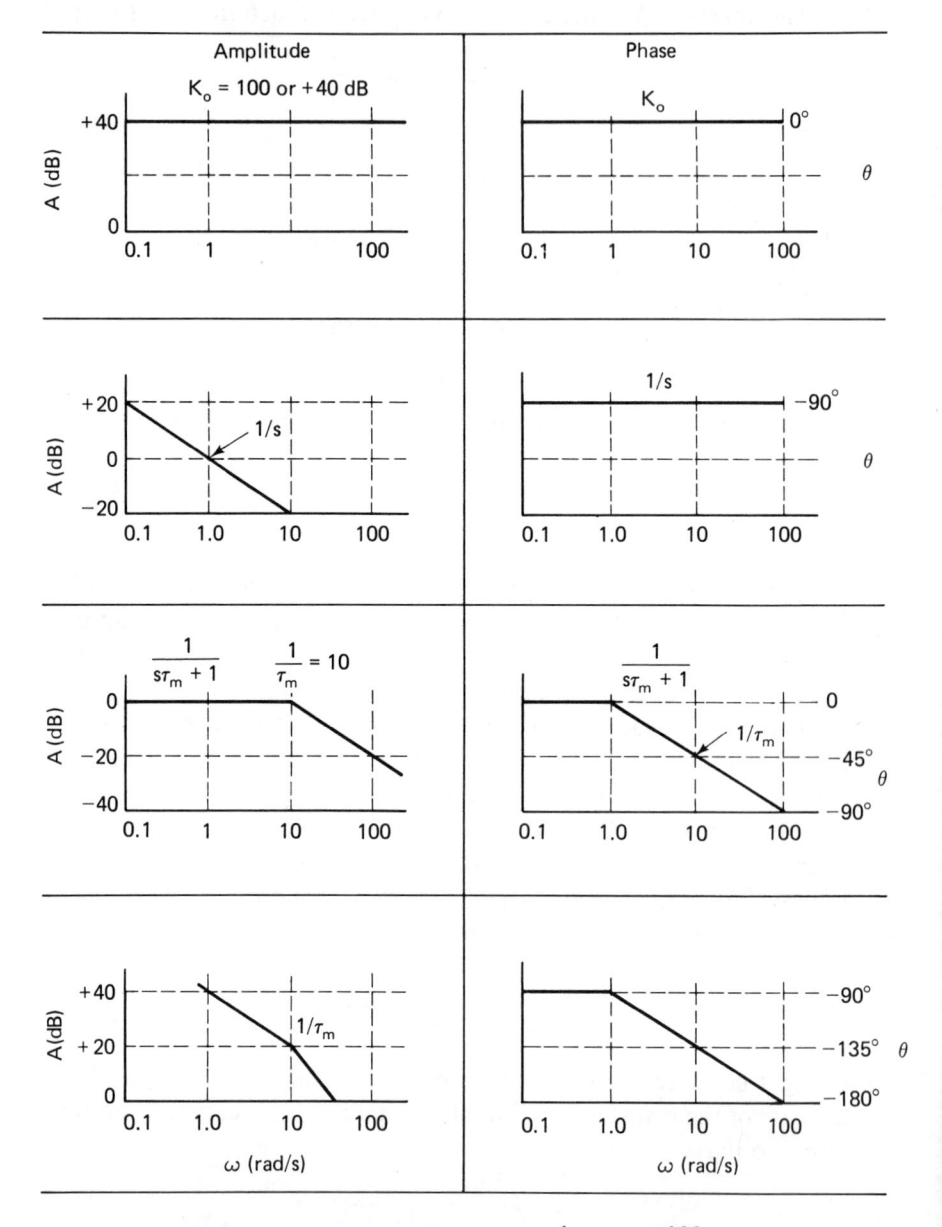

FIGURE 7-5 Bode plot of $K_o \cdot 1/s \cdot \dfrac{1}{s\tau_m + 1} = \dfrac{100}{s(0.1s + 1)}$.

Sec. 7-2 Position Control System Transfer Function **129**

For $\omega = 100$, TF $= 100/100\underline{/-90°} = 0$ dB$\underline{/-90°}$. Similarly, for $\omega = 1000$,

$$\text{TF} = \frac{100}{1000}\underline{/-90°} = \frac{1}{10}\underline{/-90°} = -20 \text{ dB}\underline{/-90°} \quad (7\text{-}12)$$

Figure 7-6 is a plot of 100/s for various values of ω.

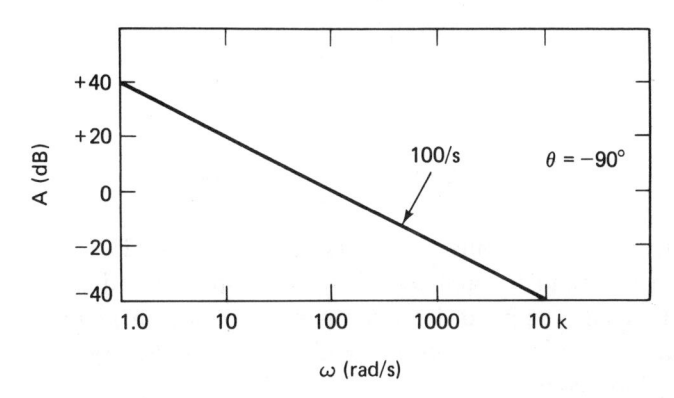

FIGURE 7-6 Bode plot of 100/s or 100·1/s.

It only remains to add the second part of the transfer function to Figure 7-6, namely, the part $1/(0.1s + 1)$, which is the plot shown in Figure 7-7.

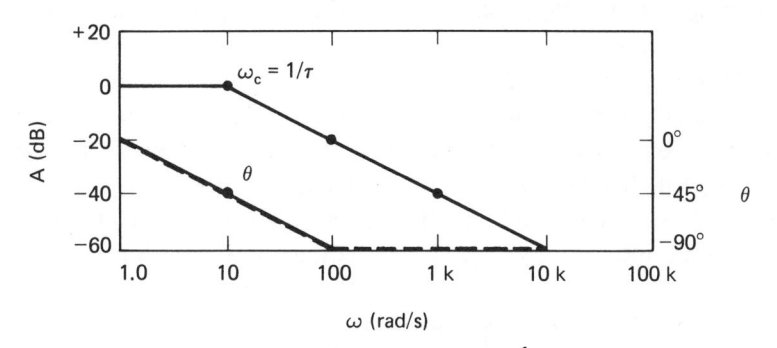

FIGURE 7-7 Bode plot of $\frac{1}{0.1s + 1}$.

The combined plot of Figure 7-6 and 7-7 becomes the plot (in solid lines) of Figure 7-8. This same plot was derived in Figure 7-5. From the graph of Figure 7-8, the gain crossover frequency of the system is $\omega_x = 30$ rad/s with a total phase shift of *approxi-*

130 Open-Loop System Analysis   Ch. 7

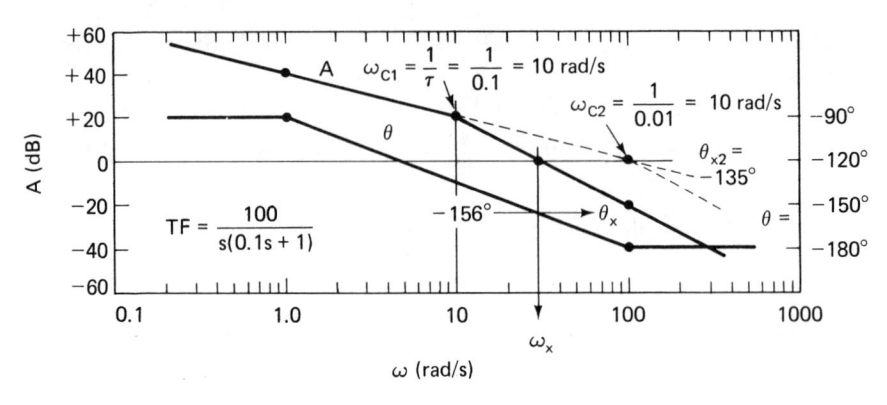

FIGURE 7-8 Bode plot of $\dfrac{100}{s(0.1s + 1)}$.

mately $-157°$ (making the phase margin only 23°). Since the system will tend to oscillate when the loop is closed and a command signal is applied, lead compensation or *plus phase* must be added to permit satisfactory performance of the system under closed-loop conditions.

7-3 GAIN EFFECTS ON SYSTEM PERFORMANCE

Let us analyze a few more complete systems by the open-loop method with the Bode plot and determine the phase margin as the overall gain is varied.

If the gain is less than 1 or 0 dB, obviously any system is stable, since it requires gain to sustain oscillations. However, it should be pointed out as a matter of interest that with low or no forward loop gain, the system will function but is "sluggish" or slow to respond to a command. In fact, friction in the follow-up potentiometer, and "stiction" or startup voltage in the motor can result in a large change in input with little or no affect on the output position. This area of no response is called *dead space* or, very often, *dead band*.

7-3-1 Dead Space

When a command signal is applied to a control system, a certain input level is needed before the output begins to respond. This level can be measured at the output of the summing junction

Sec. 7-3   Gain Effects on System Performance   131

and is the amount of input (*e*) that is required to start the system moving. In many cases, it can be as much as several volts.

Consider the system of Figure 7-3. If the input signal (*e*) must exceed ± 1 V to overcome friction, stiction, or motor startup voltage, then for an input position potentiometer sensitivity of $K_{pi} = 5$ V/rad, no output response will be obtained until the input signal exceeds ± 1 V. For the given potentiometer sensitivity, this means a position potentiometer displacement of $\pm 1/5$ rad or 22.9° total. This area of no response is the dead space.

It should be apparent that adding an amplifier ahead of the summing junction will reduce the input required (for the same amount of error), in direct proportion to the increase in the preamplifier gain. If the added gain is 10, the position input signal required to overcome the ± 1 V is

$$E_i = \pm \tfrac{1}{10} = \pm 0.1 \text{ V} \qquad (7\text{-}13)$$

In this case, the dead space for a $K_{pi} = 5$ V/rad is reduced to ± 0.02 rad or 0.04 rad total (equal to 2.3°). Mathematically,

$$\text{dead space} = \frac{\Delta e(\text{for no output response})}{K_{pa}(\text{preamp}) \times K_{pi}} \qquad (7\text{-}14)$$

EXAMPLE 7-4

If the dead space of a servo system is ± 3 V of input signal for a given $K_{pi} = 5$ V/rad, how much preamplifier gain is required to reduce the dead space to less than $\pm 1°$ or a total of 2°?

SOLUTION:

Given:

$\Delta e = \pm 3$ V or 6 V total

$K_{pi} = 5$ V/rad

K_{pa} = preamp gain required

dead space is in radians, so 2° = 2/57.3 rad

Substituting into equation (7-14),

$$\frac{2}{57.3} = \frac{6}{K_{pa} \cdot 5}$$

and

$$K_{pa} = \frac{6 \times 57.3}{2.5} = 34.38$$

However, we know that we cannot increase the system gain at random, since it can introduce instability by reducing the phase margin. Thus, an interrelationship must exist between high amplifier gain (to reduce the dead space to desirable limits) and too much gain (which can introduce instability). Since overall system gain can be modified by adding a gear reduction box with a ratio of $1/N$, then by the judicious choice of the gear reduction ratio $1/N$ and the amplifier gain K_a, both criteria can be satisfied. For the remainder of the text, to minimize the complications introduced by excessive dead space, it will be assumed that dead space is negligible and we shall proceed to evaluate system performance basically as a function of the system gain.

7-3-2 Effect of Low System Gain

Let us analyze what happens to Example 7-1 if the gain is reduced by 20 dB, or make $K_o = +20$ dB instead of $+40$ dB. The open-loop transfer function becomes

$$\text{TF} = \frac{20 \text{ dB}}{s(0.1s + 1)} \quad \text{or} \quad \frac{10}{s(0.1s + 1)} \tag{7-15}$$

The corner frequency remains at $1/\tau_m = 1/0.1 = 10$ rad/s. We will refer to the corner frequency $1/\tau_m$ as ω_m for the remainder of the text, since $1/\tau_m$ depends primarily on the motor constant; ω_c will be reserved for any electrical corner frequency other than $1/\tau_m$. The Bode plot of equation (7-15) appears as

$$\text{TF} = \frac{10}{s(0.1s + 1)}$$

where $\tau_m = 0.1$ and $\omega_m = 10$ rad/s.

A review of the graph of Figure 7-9 shows that reducing the gain of the original transfer function by 20 dB does improve the phase margin, so that it is now $180° - 135° = 45°$. This reduction in gain will provide the stable closed-loop operation. However, other secondary effects result from the lowered system gain such as a slower response to a step input and increased dead space.

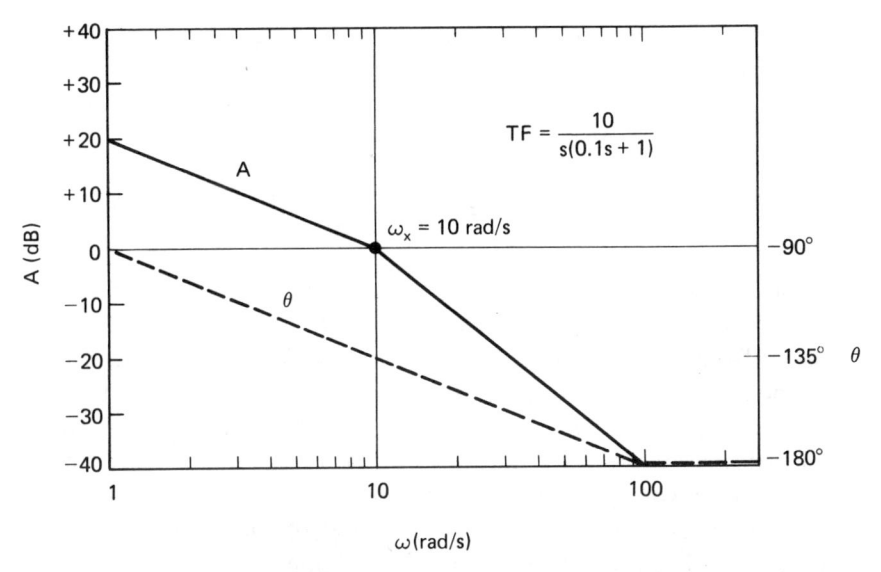

FIGURE 7-9 Effect of low system gain.

These ramifications will be discussed in more detail after we have a better understanding of closed-loop system behavior.

7-4 EFFECT OF A LEAD NETWORK ON THE PHASE MARGIN

A review of the linearized analysis of the lead network equations in Chapter 5 indicates that +45° positive phase shift is obtainable over a broad range of frequencies. By judicious choice of the corner frequencies, it should be possible to increase the phase margin at the gain crossover frequency of the system of Figure 7-8. A simple lead network that is effective is one that cancels the existing corner frequency established by $1/\tau_m$ by making the first corner of the lead network $1/\tau_m$ and the upper corner $1/10\tau_m$. In the case of Example 7-4,

$$\text{TF} = \frac{100}{s(0.1s + 1)}$$

If we make $\tau_1 = \tau_m = 0.1$ and $\tau_2 = 0.1/10 = 0.01$ s, the overall transfer function, including the lead network compensation, becomes

$$\text{TF} = \frac{1}{10} \cdot \frac{100}{s(0.1s + 1)} \frac{0.1s + 1}{0.01s + 1} = \frac{10}{s(0.01s + 1)} \quad (7\text{-}16)$$

The 20-dB loss due to the lead network must be compensated by increasing K_o to $+60$ dB by increasing K_a or reducing $1/N$ by a factor of $10x$. The new transfer function with the lead network inserted then becomes

$$\text{TF} = \frac{1000/10}{s(0.01s + 1)} \quad \text{or} \quad \frac{100}{s(0.01s + 1)} \qquad (7\text{-}17)$$

Notice in equation (7-17) that the overall K_o remains at 100 but $\omega_c = 1/0.01 = 100$ rad/s instead of 10 rad/s as originally established for $T_m = 0.1$ s. The determination of the actual corner frequencies* for the lead network is more complicated than the simple procedure shown. However, a look at Figure 7-8 and the dashed-line extension of curve A, which is the result of adding the lead network, indicates that the phase shift at ω_{x2} is $-135°$ and the phase margin is $> 40°$. This does provide the necessary phase margin for stable closed-loop operation.

For familiarization, good enough results are obtained using the lead network where the first corner frequency is the same as $1/\tau_m$. The result is a sizable increase in the permissible gain over the original condition specified. In fact, the system of equation (7-9) remains stable ($\phi > 40°$), with an increase in K_o of $+20$ dB, or 10 times.

The simple procedure of adding the lead network results in:

1. A higher effective corner frequency.
2. Permits a much higher system gain for the same phase margin, with its resultant decrease in dead space.
3. As shown in Appendix C, the system response is improved.
4. Another desirable feature occurs in the plot of Figure 7-8, namely the fact that the response at ω_{x2} is falling off at only 20 dB/decade rather than the 40 dB/decade rate as shown in the previous analysis. This point will not be stressed but is mentioned as a desirable feature, since it will

---

* A more comprehensive discussion of the lead network for phase compensation is covered in Appendix C, where a detailed analysis will show that a lead network with a corner frequency several times higher than that of the motor time constant gives a much broader range of usable gain with a phase margin of approximately 50° over a large range of K_o.

improve the system performance by lowering the peak response under steady-state conditions (details in Chapter 8).

7-5 EFFECT OF RATE FEEDBACK ON THE PHASE MARGIN

Rather than using the lead network with its $+45°$ (linearized plot or $+55°$ when corrected) positive-phase-angle, rate feedback can also be added to the feedback loop. It introduces a very interesting characteristic to the overall response. Figure 7-10 includes rate feedback in its elementary form.

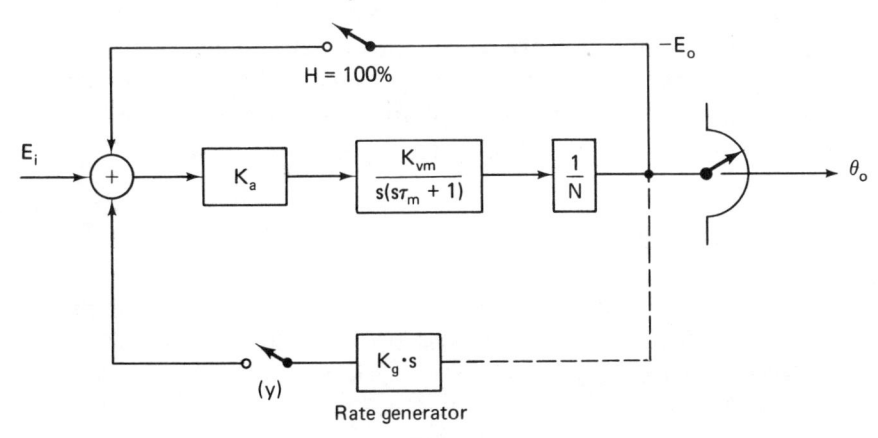

FIGURE 7-10 Position control system with rate feedback.

As shown in the section on components, the rate generator has a transfer function of $K_g \cdot s$, which makes it a differentiator. If this differentiation factor is added to the feedback loop shown, the signal (y) to the comparator is modified by $K_g \cdot s$, and the s term in the denominator is eliminated:

$$\text{TF} = \frac{K_a \cdot K_{vm} \cdot K_{po} \cdot K_g \cdot s}{s(s\tau_m + 1)} = \frac{K_a \cdot K_{vm} \cdot K_{po} \cdot K_g}{s\tau_m + 1} \qquad (7\text{-}18)$$

Equation (7-18) indicates that the phase shift cannot exceed 90° (in the frequency range of concern). The only limit to the amount of gain that can be used is the "chatter" or "jitter" that develops due to tolerances or "play" in gear trains and other coupling devices, which make it difficult at the balance to obtain an error signal near zero. Even worse, the very large phase margin* results in a sluggish response to a step input even when the higher gain is used. The addition of a voltage divider in the feedback loop makes it possible to control the amount of rate feedback until the desired response is obtained. The mathematical analysis of the effect of change in the percent of rate feedback on system behavior is a complex analysis and will not be included in this section.† The effect of rate feedback should be recognized and its use encouraged.

In Appendix B, a detailed mathematical analysis of rate feedback is presented. It shows its real adaptability for stabilizing a closed-loop system that is unstable to begin with either because of too high an open-loop gain or too much phase shift through the system. To provide the variable amount of rate feedback entails a two-loop system, one setting up the amount of rate feedback desired, and the second loop establishing the overall gain of the system when the inner and outer feedback loops are closed.

7-6 VELOCITY LAG ERROR

One final point should be mentioned before proceeding to the detailed analysis of closed-loop systems. The transfer function for the system of Example 7-5 gives

$$\text{TF} = \frac{K_a \cdot K_{vm} \cdot K_{po}/N}{s(s\tau_m + 1)} \quad (7\text{-}19)$$

and the units for the error signal become (by combining the units for K_a, K_{vm}, and K_{po} as K_o):

$$\text{TF} = \frac{E_o}{E_i} = K_o \cdot \frac{\text{V rad/s}}{\text{V}} \cdot \frac{\text{V}}{\text{rad}} = K_o \cdot \frac{1}{\text{s}} \quad (7\text{-}20)$$

Since E_o and E_i are proportional to θ_o and θ_i, respectively, K_o is referred to as the *proportionality constant* of the system (*not* the actual total loop gain).

When the system of Figure 7-4 is operated *closed loop*, it is assumed that the output shaft is rotating at the same angular

---

* See Appendix D; phase margins in excess of 60° are undesirable because of the very slow response time that results.

† Refer to Appendix B for analysis using variable rate feedback.

velocity as the input shaft. Actually, a position lag must exist between the output follow-up and input position voltages to provide the error signal (*e*) needed to drive the motor amplifier. This lag in position is referred to as the steady-state velocity lag error e_{ss} of the system.

Equation (7-4) stated that $K \cdot G = \theta_o/e$ so that

$$e_{ss} = \frac{\theta_o}{K \cdot G} \qquad (7\text{-}21)$$

Equation (7-21) shows that as *K* (the gain) increases, the error e_{ss} must decrease for the same output displacement (θ_o). There are definite limits to the amount of gain that can be used limited by instability due to excessive phase shift. As the overall system gain is increased, the gain crossover frequency is increased, and for a given corner frequency, the phase shift increases. This, of course, reduces the phase margin, which can reduce system stability.

Another undesirable effect of too much gain is a form of instability referred to as "hunting." In this case, any loss in the transfer of motion between the drive motor and the follow-up potentiometer due to gears, coupling devices, and bearings not fitting tightly (referred to as backlash) introduces a time delay in the output response. As we know, this time delay is the equivalent of additional phase lag and the result is a decrease in the phase margin. When there is sufficient gain to produce oscillations due to the backlash, it is called "hunting" and is characterized by a nonlinear form of oscillation, a sort of jumping back and forth. K_o can be easily controlled by adding a gain control in the preamplifier. In the analyses that follow, the velocity lag error has little affect on system performance and will be neglected.

GLOSSARY

Dead Space: The incremental input for which there is no response in output; it is directly related to the start up voltage of the motor.

Jitter: A chatter of the mechanical output device, usually a shaft output or position indicator due to excessive gain in the system.

Open Loop Analysis: The establishing of the phase margin at the gain crossover frequency by use of the Bode plot after a system equation has been determined.

Phase Compensation: The adding of a lead network or rate feedback to increase the phase margin at the gain crossover frequency.

Position Control System: A closed-loop system wherein the mechanical output (usually angular position θ_o) is directly proportional to the input position (also an angular position θ_i).

Proportionality Constant: K_o is the ratio of the input to the comparator to the output fed back to the comparator. It includes all elements and subsystems that affect the feedback signal to the comparator.

Rate Feedback: Sometimes also referred to as "velocity feedback." The mechanical output (angular position) feeds a tachometer whose output is proportional to the change in output speed. The output voltage, which becomes part of the feedback, is altered by the transfer function of the tachometer ($K_g \cdot s$). This adds a positive phase angle to the feedback signal, increasing the phase margin.

Velocity Lag Error: The lag between the output rotational response and the input command, which develops an error signal e_{ss} necessary to drive the servo motor. The higher the system gain, the lower is the velocity lag error.

ω_m: The corner frequency of the motor $1/\tau_m$.

PROBLEMS

7.1 What is the velocity constant K_{vm} of a motor in rad/s/V when given as 5000 rpm at rated input of 12 V?

7.2 What is the sensitivity of a position potentiometer that has a total rotational range of 150° when it is energized by a ± 12-V source? Express your answer in V/rad.

7.3 The position potentiometer of Problem 7.2 is used to drive a position control servo. However, it was found that no change in output occurred until the input voltage was ± 0.25 V.

  (a) What is the dead space in degrees?

  (b) How much must the system gain be increased to reduce the dead space to less than $\pm 0.25°$?

7.4 Given the transfer function TF $= 10/s(0.5s + 1)$, what is the phase margin at the gain crossover frequency? (Use the Bode plot.)

7.5 Given the transfer function TF $= 30/s(0.5s + 1)$, determine the phase margin at the gain crossover frequency. (Use the Bode plot.)

7.6 In the equation given in Problem 7.5, replace s by $j\omega$ and substitute the gain crossover frequency found in the Bode plot solution. Does the amplitude and phase at this frequency agree with the answer found in Problem 7.5?

7.7 Given:

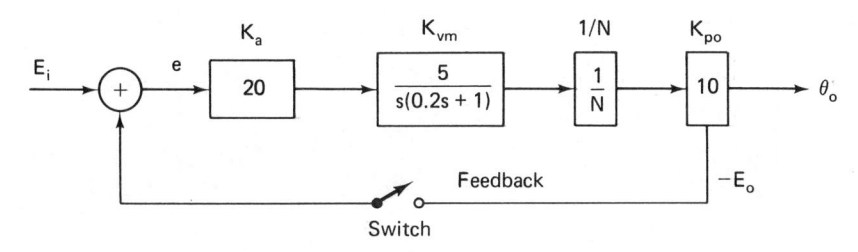

  (a) What is the phase margin ϕ_x (open loop) when $N = 20$? (TF $= E_o/e$ and $e = E_i$.)

  (b) When $N = 50$.

  (c) What happens to the phase margin as the system gain is decreased?

  (d) Will the systems be stable under closed loop operation?

7.8 (a) In the system of Problem 7.7 with $N = 20$, add a simple lead network to raise the corner frequency of the overall transfer function by $10\times$. What happens to the phase margin?

  (b) Increase the gain by $10\times$ (the low-frequency attenuation characteristic of the lead network) and determine the slope of the output and the phase margin at the gain crossover frequency.

(c) Is part (a) stable for closed-loop operation?

7.9 Determine the torque constant K_{tm} of a motor which can produce a maximum torque of 20 in.-oz when driven by an input of 12 V.

7.10 Given the ac control system shown where $\theta_o/\theta_i = 1$:

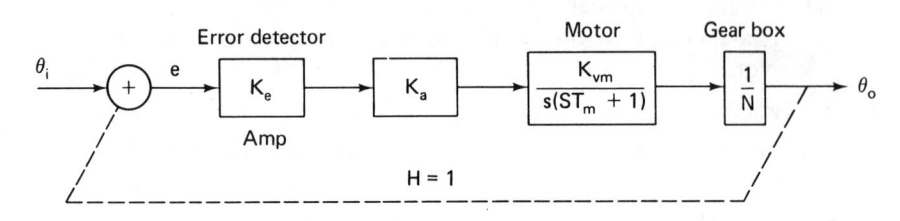

(a) Write the open loop transfer function in terms of θ_o/e given:

$$K_e = 8 \text{ V/radian}$$
$$K_a = 10$$
$$K_{vm} = 5/s(0.3s + 1)$$
$$N = 20$$

8

Closed-Loop System Analysis

8-1 INTRODUCTION

Closed loop feedback systems are divided into three basic types, referred to as the Type 0, Type 1, or Type 2 system. It is interesting to note that the classification is based on the number of integrations or $1/s$ that appear in the open-loop transfer function. The basic position control system equation is of the form

$$\text{TF} = \frac{K}{s(s\tau_m + 1)} \quad \text{or} \quad \frac{1}{s} \cdot \frac{K}{(s\tau_m + 1)} \tag{8-1}$$

Since it contains one integration $(1/s)$, it is called a *Type 1 system*. In the case of the speed control system,* the transfer function was shown to be

$$\text{TF} = \frac{K}{s\tau_m + 1} \tag{8-2}$$

* Covered separately in Appendix E.

141

and no $1/s$ is present. Thus, the speed control system is a *Type 0 system*.

Special emphasis will be placed on the Type 1 system since the basic position control system or servo fits into this category.

The closed-loop analysis will be done in several ways:

1. By the use of complex algebra.
2. By a graphical method called the Nichols chart.
3. By transient analysis, the use of a step function $1/s$ as the input.

8-2 CLOSED-LOOP-POSITION CONTROL SYSTEM

In control system analysis, the forward loop gain is presented in a complex manner. In addition to the dc amplitude characteristics, the automatic control system closed-loop equation also includes the effect of frequency on the gain of the system.

Figure 8-1 shows the block diagrams of a conventional voltage amplifier with feedback and an automatic control system with feedback. The closed-loop gain equations remain unchanged

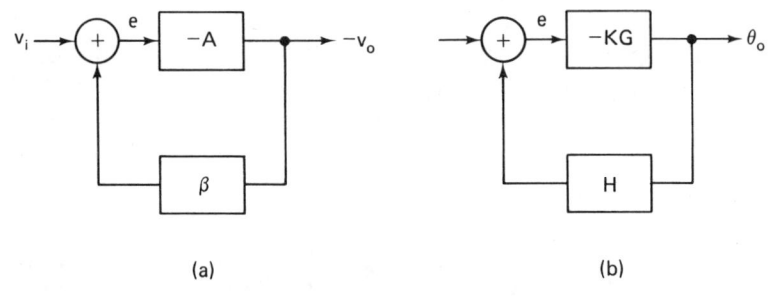

FIGURE 8-1 (a) Voltage amplifier with negative feedback; (b) automatic control system with negative feedback.

except for the new block designations. Note that A becomes KG, where K is the forward loop gain and G a function of s; β becomes $H = \%$ negative feedback.

The closed-loop gain of Figure 8-1(a), with the designations

Sec. 8-2   Closed-Loop-Position Control System   143

for control system analysis shown in Figure 8-1(b), changes:

$$G = \frac{A}{1 + A\beta} \quad \text{to} \quad \frac{\theta_o}{\theta_i} = \frac{KG}{1 + KGH} \tag{8-3}$$

and for $H = 1$ or 100% feedback:

$$\frac{\theta_o}{\theta_i} = \text{TF}_{\text{CL}} = \frac{KG}{1 + KG} \tag{8-4}$$

8-2-1   Adding a Gear Train

When a gear train is added to the output to increase the output torque or to lower the system gain KG, the forward loop gain is modified by the gear reduction ratio $1/N$ and the complete closed-loop equation with $H = 1$ becomes

$$\frac{\theta_o}{\theta_i} = \text{TF}_{\text{CL}} = \frac{KG/N}{1 + KG/N} \tag{8-5}$$

and for $K_o = K/N$ as has been the case:

$$\text{TF}_{\text{CL}} = \frac{K_o \cdot G}{1 + K_o \cdot G} \tag{8-6}$$

For the typical *position control system equations* we have been analyzing, the open-loop gain is given as

$$\text{TF}_{\text{OL}} = K_o \cdot G = \frac{K_o}{s(s\tau_m + 1)} \tag{8-7}$$

The closed-loop gain [substituting equation (8-7)] into equation (8-6) becomes

$$\text{TF}_{\text{CL}} = \frac{K_o \cdot \dfrac{1}{s(s\tau_m + 1)}}{\left[1 + K_o \cdot \dfrac{1}{s\tau_m + 1}\right]} \frac{s(s\tau_m + 1)}{s(s\tau_m + 1)} \tag{8-8}$$

$$= \frac{K_o}{s(s\tau_m + 1) + K_o}$$

8-3 STEADY-STATE ANALYSIS

In Chapter 7, it was shown that too high a system gain resulted in too low a phase margin. By reducing the system gain, the phase margin was increased to 45°, which provides for stable closed-loop performance.

In the evaluation of closed-loop performance, we will deliberately begin with a system that has too low a proportionality constant K_o. Later, the gain will be increased and the closed-loop response will again be evaluated.

Given

$$\text{TF}_{\text{OL}} = \frac{K_o}{s(s\tau_m + 1)} = \frac{10}{s(0.05s + 1)} \quad (8\text{-}9)$$

for which the open-loop Bode plot is shown in Figure 8-2. (Note the phase margin at ω_x is approximately 58°.) The closed-loop transfer function for equation (8-9) becomes

$$\text{TF}_{\text{CL}} = \frac{K_o}{s(s\tau_m + 1) + K_o} = \frac{10}{s(0.05s + 1) + 10} \quad (8\text{-}10)$$

A point-by-point plot of the closed-loop response is easily accom-

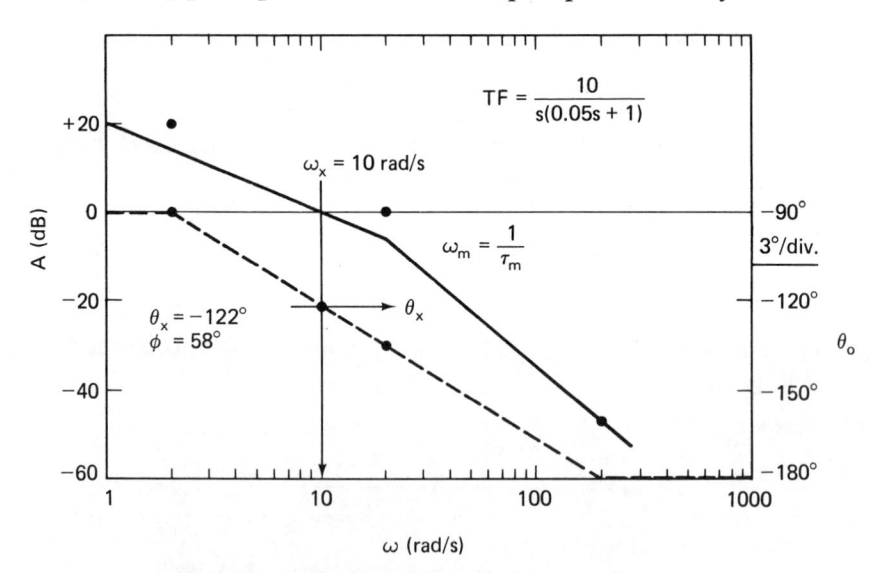

FIGURE 8-2 Open-loop Bode plot of $10/[s(.05s + 1)]$.

Sec. 8-3 Steady-State Analysis 145

plished by replacing all s operators in equation (8-10) by $j\omega$ and solving for $A/\underline{\theta}$ for various values of ω above and below the corner frequency $1/\tau_m = \omega_m$.

$$\text{TF}_{\text{CL}} = \frac{10}{j\omega(0.05j\omega + 1) + 10} \quad \text{or} \quad \frac{10}{-0.05\omega^2 + j\omega + 10} \quad (8\text{-}11)$$

Substituting values of ω around $\omega_m = 1/0.05 = 20$ rad/s, as for example $\omega = 1, 5, 10, 20, 50, 100,$ and 200 will provide $A/\underline{\theta}$ information for a point-by-point amplitude and phase plot of the closed-loop system under consideration. Several solutions* of $A/\underline{\theta}$ versus ω are presented:
For $\omega = 1$,

$$A/\underline{\theta} = \frac{10}{-(0.05)(1)^2 + j1 + 10} = \frac{10}{9.95 + j1}$$
$$= \frac{10}{\sqrt{(9.95)^2 + 1^2}/\tan^{-1}(1/9.95)} = \frac{10}{10}/\underline{-5.7°} = 0 \text{ dB}/\underline{-5.7°}$$

For $\omega = 20$,

$$A/\underline{\theta} = \frac{10}{-(0.05)(20)^2 + j20 + 10} = \frac{10}{-10 + j20}$$
$$= -7 \text{ dB}/\underline{-116°}$$

For $\omega = 50$,

$$A/\underline{\theta} = \frac{10}{-(0.05)(50)^2 + j50 + 10} = \frac{10}{-115 + j50}$$
$$= -22 \text{ dB}/\underline{-156°}$$

The procedure is extended to cover a number of other frequencies, which are summarized in Table 8-1. The Bode plot of the

---

* An electronic calculator is most convenient for the solution of $A/\underline{\theta}$ versus ω. By rewriting equation (8-11) as

$$A/\underline{\theta} = \left(\frac{1}{(1 - 0.005\omega^2) + j0.1\omega}\right)$$

a programmable calculator will provide $A/\underline{\theta}$ directly as ω is varied.

closed-loop system of equation (8-10) are the data of Table 8-1 shown in Figure 8-3.

TABLE 8-1 $A/\underline{\theta}$ versus ω for the system of equation (8-10).

ω	A (dB)	θ (deg)
1.0	0	−5.7
2.0	0	−11.5
3.0	0	−17.4
4.0	−0.03	−23.5
5.0	−0.07	−29.7
7.0	−0.25	−42.8
10.0	−0.97	−63.4
15.0	−3.55	−94.8
20.0	−7	−116
50.0	−22	−157
100	−34	−168
200	−46.4	−175
400	−58	−177

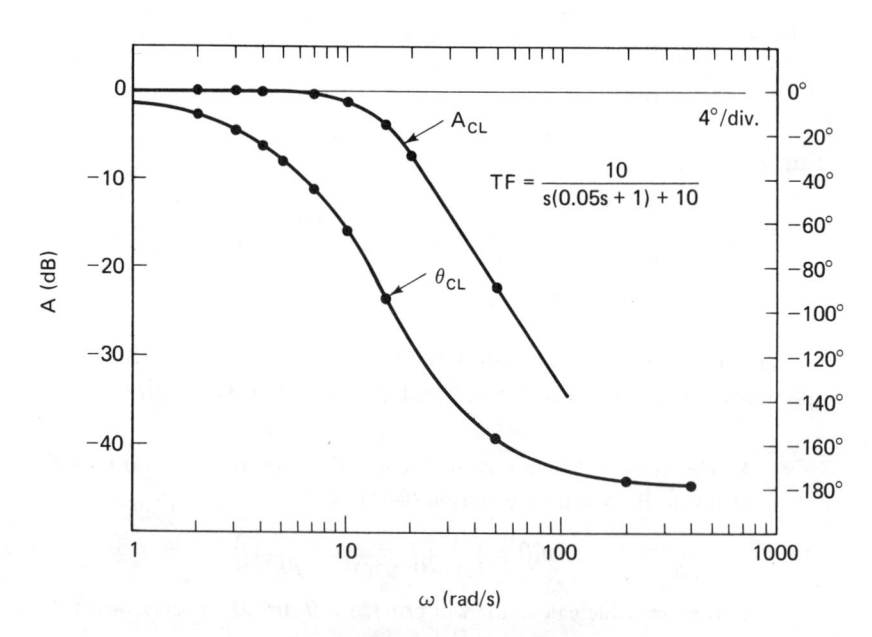

FIGURE 8-3 Plot of Table 8-1.

8-4 EFFECT OF GAIN ON CLOSED-LOOP RESPONSE

In the preceding development of the closed-loop response, if the gear train is replaced with one of 2:1 reduction instead of 20:1 as was used, a considerable change occurs in the closed-loop response. The modified transfer function with a 10× increase in K_o becomes

$$\text{TF}_{\text{CL}} = \frac{K_o}{s(s\tau_m + 1) + K_o} = \frac{100}{s(0.05s + 1) + 100}$$

and

$$\text{TF}_{\text{CL}} = \frac{100}{-0.05\omega^2 + j\omega + 100} \qquad (8\text{-}12)$$

Using the procedure of substituting different values of ω in equation (8-12), we obtain Table 8-2. The plot of Table 8-2 is shown in Figure 8-4.

TABLE 8-2 A/θ versus ω for system equation (8-12).

ω	A (dB)	θ (deg)
1.0	0	−0.6
3.0	0	−1.8
5.0	+0.1	−2.87
10	+0.4	−6
15	+0.92	−9.6
20	+1.67	−14
30	+4.1	−28.6
40	+7.00	−63.4
50	+5.0	−116.6
100	−12.4	−167
200	−25.6	−174

8-4-1 Peak Response M_p at ω_p

An analysis of Figure 8-3 shows that with the gear ratio of 20:1, the K_o is not adequate for fast response. In fact, the output at the corner frequency of the motor, 1/0.05 or 20 rad/s, is down

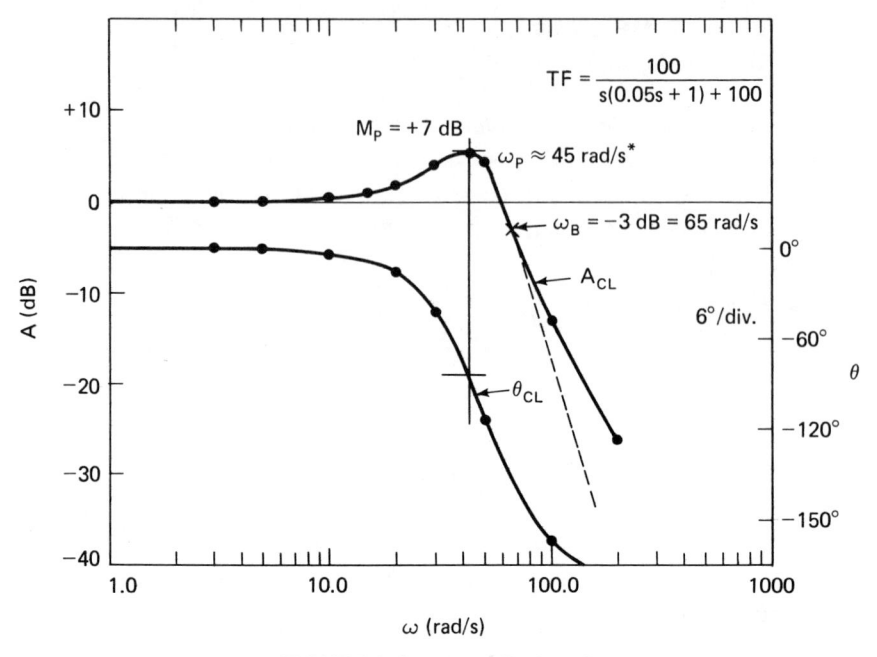

FIGURE 8-4 Plot of Table 8-2.

by some −7 dB. By increasing the K_o to 100, either by reducing the gear ratio to 10:1 or increasing the amplifier gain by 10×, a *resonant frequency* ω_p with a peak response M_p of +7 dB occurs at about 45 rad/s (clearly seen in Figure 8-4).

Again, an electrical/mechanical analogy will clearly show the reason for the phenomena. In electrical resonant systems, two basic conclusions commonly recognized are:

1. There is a resonant condition which is based on the premise that $X_L = X_C$, producing a unity power factor, in which case the output current is in phase with the input voltage. The frequency is the natural resonant frequency (ω_n) and is based on $\omega L = 1/\omega C$, from which

$$\omega_n = \frac{1}{\sqrt{LC}} \qquad (8\text{-}13)$$

2. When losses are present (theoretically, an inductance will always have losses (R)), the maximum voltage output across the capacitor *does not* occur when $X_L = X_C$ but rather occurs at a slightly lower frequency. What actually happens is that

Sec. 8-5 *The Damping Factor* **149**

as the input frequency is decreased, the reactance of the capacitor increases slightly faster than the current decreases. The cross point where $I \cdot X_C$ is maximum is a solution using the calculus* and in terms of the losses is found to be

$$\omega_p = \omega_n \sqrt{1 - 2\zeta^2} \qquad (8\text{-}14)*$$

where $\zeta = 1/2Q =$ damping factor
$Q = \omega L/R$

8-5 THE DAMPING FACTOR

In Chapter 4, we mentioned that electrical resonant systems are affected by losses present (primarily in the coil). The effects were summarized by equation (4-12), which is repeated here:

$$i(t) = \frac{E}{\omega L} \cdot e^{-at} \sin \omega \cdot t \qquad (8\text{-}15)$$

where $a = R/2L$, and the rate of decay of the sine wave depends on the decaying exponential e^{-at}. This is the part of the equation showing the rate at which energy is dissipated in a resonant circuit as a function of time. Each time reactive power converts from vars inductive to vars capacitive, the current is reduced as power is dissipated in the resistive losses (R). The higher the input frequency, the faster is the rate of decay for a given damping coefficient.

Mathematically, the damping factor ζ, which describes the rate of decay in terms of rad/s rather than in time only, becomes

$$\zeta = \frac{a}{\omega}$$

where $a = R/2L$
$\omega =$ resonant frequency

or

$$\zeta = \frac{R}{2L} \frac{1}{\omega} = \frac{R}{2\omega L} \qquad (8\text{-}16)$$

Since

$$Q = \frac{\omega L}{R} \text{ then } \zeta = \frac{1}{2Q} \qquad (8\text{-}17)$$

* Derivation using the calculus is given by equation (8-32).

150 Closed-Loop System Analysis   Ch. 8

Note that equation (8-16) contains an R/L relationship and since $L/R = \tau$, R/L must equal $1/\tau$. Equation (8-17) can be rewritten in a different form* as

$$\zeta = \frac{1}{2\omega_n \tau} \qquad (8\text{-}18)*$$

where $\omega_n = 1/\sqrt{L \cdot C}$
         $\tau = $ time constant L/R

8-5-1 Natural Resonance ω_n

Statement number (1) in Section 8-4-1 specifies that the current in a series resonant circuit is exactly in phase with the input voltage, in which case the output voltage $v_o = i(jX_C) = A\underline{/-90°}$. Thus, at the natural resonant frequency (ω_n), the output voltage (v_o) is shifted $-90°$ with respect to the input voltage (v_i).

A look at equation (8-7), written with $j\omega$ in place of s, results in the transfer function:

$$\text{TF} = \frac{K_o}{s(s\tau_m + 1) + K_o} = \frac{K_o}{j\omega(j\omega\tau_m + 1) + K_o} \qquad (8\text{-}19)$$

$$= \frac{K_o}{j^2\omega^2\tau_m + j\omega + K_o}$$

and for $j^2 = -1$,

$$A\underline{/\theta} = \frac{K_o}{-\omega^2\tau_m + j\omega + K_o} \qquad (8\text{-}20)$$

In order for the output of equation (8-20) to result in an $A\underline{/\theta} = A\underline{/-90°}$ (which is the case at natural resonance), the denominator must be reduced so only the $j\omega$ or the $\omega\underline{/-90°}$ term remains. This occurs when $K_o - \omega^2\tau_m$ is made equal to zero, from which

$$\omega^2 = \frac{K_o}{\tau_m} \quad \text{and} \quad \omega = \sqrt{\frac{K_o}{\tau_m}} \qquad (8\text{-}21)$$

Equation (8-21) satisfies the conditions for natural resonance, resulting in

$$\omega = \omega_n = \sqrt{\frac{K_o}{\tau_m}} \qquad (8\text{-}22)$$

---

* This same definition for ζ is derived in Chapter 10, using the Laplace transformation.

Sec. 8-5   The Damping Factor   151

EXAMPLE 8-1

For the closed-loop system of equation (8-12), determine the damping factor ζ.

SOLUTION:

Use equation (8-18):

$$\zeta = \frac{1}{2\omega_n \tau_m}$$

where

$$\omega_n = \sqrt{\frac{K_o}{\tau_m}} = \sqrt{\frac{100}{0.05}} = 44.7 \text{ rad/s}$$

making

$$\zeta = \frac{1}{2(44.7)(0.05)} = 0.224$$

8-5-2 Determining the Peak Frequency Response ω_p

Equation (8-20) can be modified to provide the peak frequency response (ω_p) for any value of damping factor. Dividing the numerator and denominator of equation (8-20) by τ_m results in

$$A\underline{/\theta} = \frac{K_o/\tau_m}{-\omega^2 + j\omega/\tau_m + K_o/\tau_m} \qquad (8\text{-}23)$$

Substituting ω_n^2 for K_o/τ_m, equation (8-23) becomes

$$A\underline{/\theta} = \frac{\omega_n^2}{\omega_n^2 + j\omega/\tau_m - \omega^2} \qquad (8\text{-}24)$$

Returning to equation (8-18), where

$$\zeta = \frac{1}{2\omega_n \tau_m}, \quad \frac{1}{\tau_m} = 2\zeta\omega_n \qquad (8\text{-}25)$$

equation (8-24) can be rewritten to provide the full amplitude and phase characteristics of any oscillatory system as the damping factor is varied. Substituting equation (8-25) into equation (8-24) gives

$$A\underline{/\theta} = \frac{\omega_n^2}{\omega_n^2 + j\omega \cdot 2\zeta\omega_n - \omega^2} \qquad (8\text{-}26)$$

152 Closed-Loop System Analysis Ch. 8

Dividing the numerator and denominator by ω_n^2,

$$\underline{A/\theta} = \frac{1}{1 + j2\zeta(\omega/\omega_n) - (\omega/\omega_n)^2} \tag{8-27}$$

If we normalize equation (8-27) by making $\omega/\omega_n = 1$ (as we did in our Bode plot presentations), a universal plot can be prepared of $\underline{A/\theta}$ for any resonant frequency by establishing the multiplier n. To make a good plot, a number of frequencies above and below $(\omega/\omega_n) = 1$ are used. Several solutions of $\underline{A/\theta}$ are given for familiarization as ω and ζ are varied.

CASE 1

Determine $\underline{A/\theta}$ for $\omega/\omega_n = 1$ with $\zeta = 0.1$. Substituting into equation (8-27),

$$\underline{A/\theta} = \frac{1}{1 + j2(0.1)(1) - 1} = \frac{-j}{0.2} = +14 \text{ dB}\underline{/-90°}$$

CASE 2

Given $\omega/\omega_n = 0.707$ and $\zeta = 0.3$,

$$\underline{A/\theta} = \frac{1}{1 + j2(0.3)(0.707) - (0.707)^2} = \frac{1}{1 + j0.42 - 0.5}$$
$$= +5.7 \text{ dB}\underline{/-40°}$$

CASE 3

Given $\omega/\omega_n = 2.0$ and $\zeta = 0.707$,

$$\underline{A/\theta} = \frac{1}{1 + j2(0.707)(2) - 2^2} = \frac{1}{j2.83 - 3}$$

Vectorially:

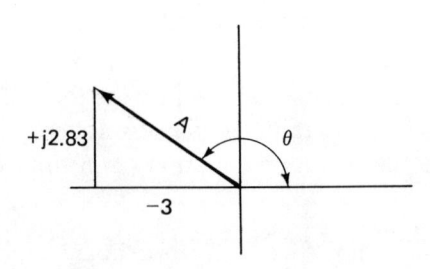

Sec. 8-5   The Damping Factor   153

$$A\underline{/\theta} = \frac{1}{4.2\underline{/+136.67°}} = -12.3 \text{ dB}\underline{/-136.67°}$$

Figure 8-5 is the composite of $A\underline{/\theta}$ for values of ζ from 0.05 to 1 from one decade below to one decade above $\omega/\omega_n = 1$. As can be seen from the graphs of Figure 8-5, as the damping increases, the

$$A\underline{/\theta} = \frac{1}{1 + j2Z(\omega/\omega_n) - (\omega/\omega_n)^2} = \frac{1}{1 + j2Z\omega - \omega^2}$$

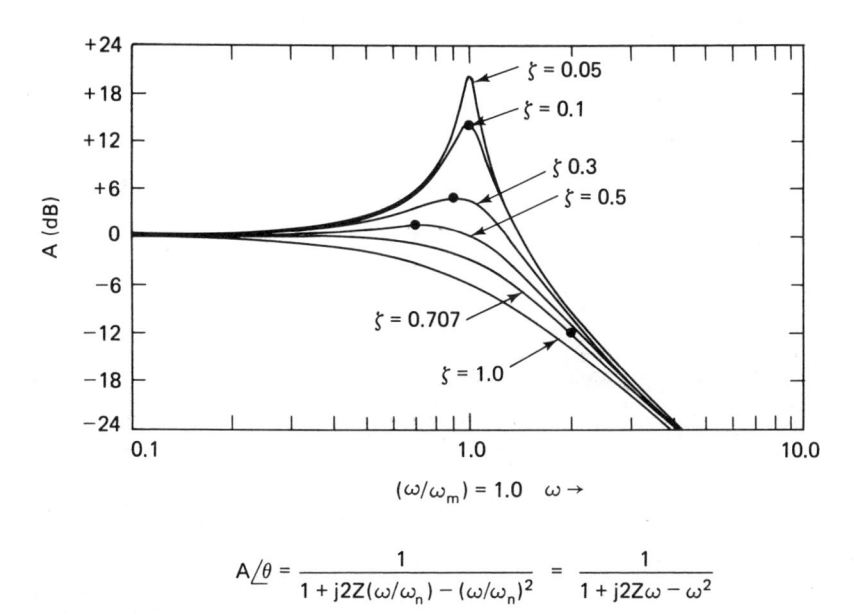

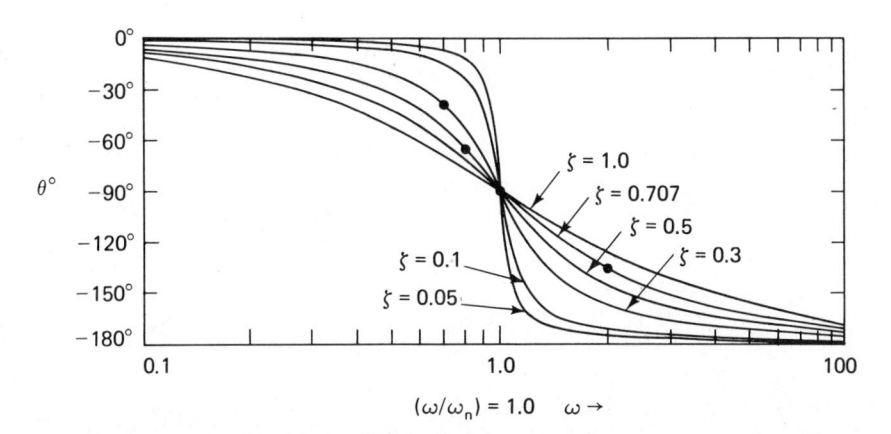

FIGURE 8-5 Amplitude and phase curves of a closed-loop system under steady state conditions as ζ is varied. Points shown correspond to Case I & II.

154 Closed-Loop System Analysis   Ch. 8

phase shift at ω_n remains constant (at $-90°$), but the peak frequency ω_p and its phase shift begin to decrease. Once ζ has been determined, $\omega_p{}^*$ is found by substituting in equation (8-28):

$$\omega_p = \omega_n\sqrt{1 - 2\zeta^2} \qquad (8\text{-}28)$$

EXAMPLE 8-2

Let us verify that equation (8-28) provides ω_p in terms of ω_n as ζ is varied.

SOLUTION:

CASE 1

For $\zeta = 0.3$, $\omega_p \approx 0.9\omega_n$ from the graph of Figure 8-5.

$$\omega_p = \omega_n\sqrt{1 - 2(.3)^2} = \omega_n\sqrt{0.82} = 0.9\omega_n$$

* Derivation of Equation (8-28)

Equation (8-27) for $\omega_n = 1$ is simplified to

$$A\underline{/\theta} = \frac{1}{1 + j2\zeta\omega - \omega^2} \qquad (8\text{-}29)$$

which can be rewritten by combining terms as

$$1/A\underline{/\theta} = (1 - \omega^2) + j2\zeta\omega$$

and

$$1/A\underline{/\theta} = \sqrt{(1 - \omega^2)^2 + (2\zeta\omega)^2}$$

so that

$$1/A\underline{/\theta} = \sqrt{1 - 2\omega^2 + \omega^4 + 4\zeta^2\omega^2} \qquad (8\text{-}30)$$

To obtain the maximum amplitude of A for a frequency requires differentiation and setting the result equal to zero, so

$$0 - 4\omega + 4\omega^3 + 2\omega(4\zeta^2) = 0 \qquad (8\text{-}31)$$

Dividing equation (8-31) by 4ω results in

$$-1 + \omega_p^2 + 2\zeta^2 = 0 \quad \text{and} \quad \omega_p^2 = 1 - 2\zeta^2$$

In terms of ω_n, $\omega_p^2/\omega_n^2 = 1 - 2\zeta^2$ from which

$$\omega_p = \omega_n\sqrt{1 - 2\zeta^2} \qquad (8\text{-}32)$$

Sec. 8-5   The Damping Factor   **155**

CASE 2

For $\zeta = 0.5$ and $\omega/\omega_n = 1$, $\omega_p = 0.7\omega_n$.

$$\omega_p = \omega_n\sqrt{1 - 2(0.5)^2} = \omega_n\sqrt{1 - 0.5} = \omega_n\sqrt{0.5} = 0.707\omega_n$$

EXAMPLE 8-3

For the system under consideration, with $\tau_m = 0.05$ s and $K_o = 100$, determine ω_p and M_p.

SOLUTION:

1st
$$\omega_n = \sqrt{K_o/\tau_m} = \sqrt{100/0.05} = 44.72 \text{ rad/s}$$

2nd
$$\zeta = 1/2\omega_n\tau_m = 1/2 \times 44.72 \times 0.05 = 0.224$$

3rd
$$\omega_p = \omega_n\sqrt{1 - 2\zeta^2}$$
$$= 44.72\sqrt{1 - 2(0.224)^2}$$
$$= 42.00 \text{ rad/s}$$

Using equation (8-12) with

$$A\underline{/\theta} = \frac{100}{-0.05(42)^2 + j42 + 100}$$
$$= \frac{100}{-88.2 + j42 + 100}$$
$$= \frac{100}{11.8 + j42} = 2.29\underline{/-74.3°} \quad (8\text{-}33)$$
$$= +7.2 \text{ dB}\underline{/-74.3°}$$

This agrees fairly well with the results shown in the point-by-point plot of Figure 8-4.

8-5-3 Determining the Bandwidth (ω_B) of a Closed Loop System

Bandwidth is defined as the range of frequencies to which a system will respond before the output begins to fall off (taken as 0.707, $1/\sqrt{2}$ or -3 dB of its midfrequency response). The bandwidth requirements of a system depend on the application of the intended system and is part of the system design specification. However, the bandwidth (ω_B) of a system can be found readily when the damping factor and the natural frequency (ω_n) have been determined. Equation (8-27) solves for system gain as a function of frequency ω. By setting the gain ($A/\underline{\theta}$ for -3 dB or $1/\sqrt{2}$, the frequency that satisfies this condition must be ω_B, the bandwidth. Therefore, again using equation (8-27),

$$A/\underline{\theta} = \frac{1}{1 + j2\zeta(\omega_B/\omega_n) - (\omega_B/\omega_n)^2} = \frac{1}{\sqrt{2}} \qquad (8\text{-}34)$$

For ease in handling equation (8-34), let us replace ω_s/ω_n by ω; then

$$A/\underline{\theta} = \frac{1}{\sqrt{2}} = \frac{1}{1 + j(2\zeta\omega) - \omega^2} \qquad (8\text{-}35)$$

Rearranging terms in the denominator $(1 + j2\zeta\omega - \omega^2)$ to $(1 - \omega^2) + j2\zeta\omega$ gives us

$$\sqrt{2} = (1 - \omega^2) + j2\zeta\omega \qquad (8\text{-}36)$$

Vectorially, then, equation (8-36) appears as

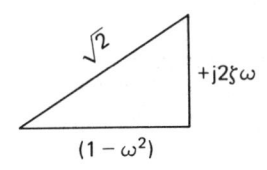

Using the Pythagorean theorem, $c^2 = a^2 + b^2$,

$$2 = (2\zeta\omega)^2 + (1 - \omega^2)^2$$

we obtain

$$2 = 4\zeta^2\omega^2 + (1 - 2\omega^2 + \omega^4)$$

which simplifies to

$$\omega^4 - \omega^2(2 - 4\zeta^2) - 1 = 0 \qquad (8\text{-}37)$$

A variation of the use of the quadratic equation simplifies the

Sec. 8-5   The Damping Factor   **157**

solution of equation (8-37) for values of ω as ζ is varied. Equation (8-37) can be rewritten as

$$\omega^{2n} - \omega^{n}(2 - 4\zeta^{2}) - 1 = 0 \quad \text{with } n = 2 \quad (8\text{-}38)$$

The solution for ω is handily accomplished by making

$$\omega^{n} = \frac{-b \pm \sqrt{b^{2} - 4ac}}{2a} \quad (8\text{-}39)$$

From equation (8-38),

$$n = 2$$
$$a = 1$$
$$b = -(2 - 4\zeta^{2})$$
$$c = -1$$

EXAMPLE 8-4

Solve for ω with $\zeta = 0.1$ (recall that $\omega = \omega_{B}/\omega_{n}$).

SOLUTION:

Substituting values into equation (8-39) with

$$n = 2$$
$$a = 1$$
$$b = -(2 - 4\zeta^{2}) = -[2 - 4(0.1)^{2}] = -1.96$$
$$c = -1$$

results in

$$\omega^{2} = \frac{1.96 \pm \sqrt{3.84 + 4}}{2} \quad \text{or} \quad \omega^{2} = 2.38$$

and

$$\omega = \sqrt{2.38} = 1.543$$

and for $\omega = \omega_{B}/\omega_{n}$:

$$\omega = \frac{\omega_{B}}{\omega_{n}} = 1.543 \quad \text{and} \quad \omega_{B} = 1.543\omega_{n}$$

EXAMPLE 8-5

Solve for ω_B/ω_n for $\zeta = 0.2$.

SOLUTION:

Using equation (8-39) with

$$n = 2$$
$$a = 1$$
$$b = -[2 - 4(0.2)^2] = -1.84$$
$$c = -1$$

$$\omega^2 = \frac{1.84 \pm \sqrt{3.386 + 4}}{2} = 2.28 \quad \text{and} \quad \omega = 1.51$$

from which

$$\omega_B = 1.51\omega_n$$

Table 8-3 is a compilation of ω_B/ω_n for a number of values of ζ. Note in equation (8-39) that an equation for ω_B/ω_n in general terms can readily be developed using the quadratic equation with

$$n = 2$$
$$a = 1$$
$$b = -(2 - 4Z^2)$$
$$c = -1$$

Since

$$\omega^n = \frac{-b \pm \sqrt{b^2 - 4ac}}{2a} \quad \text{or} \quad \left(\frac{\omega_B}{\omega_n}\right)^2 = \frac{-b \pm \sqrt{b^2 - 4ac}}{2a} \quad (8\text{-}40)$$

After substituting a, b, and c into equation (8-40),

$$\omega_B = \omega_n \sqrt{\frac{(2 - 4\zeta)^2 + \sqrt{(2 - 4\zeta)^2 + 4}}{2}} \quad (8\text{-}41)$$
$$= \omega_n \sqrt{1 - 2\zeta^2 + \sqrt{2 - 4\zeta^2 + 4\zeta^4}}$$

Since $\omega_n = \sqrt{K_o/\tau_m}$, we can now prepare a handy guide (Table 8-3) from which we can quickly estimate ω_p, M_p, and ω_B for a specific damping factor (ζ). In Chapter 10, it will be shown that a ζ between 0.4 and 0.5 will provide good system response. Figure

8-5 indicates this will result in a maximum amplitude (M_p) at (ω_p) of approximately +2 dB.

TABLE 8-3 A/θ response of a closed-loop system as the damping factor ζ is varied.

ζ	ω_p/ω_n	M_p (dB)	ω_B/ω_n
0.05	1	+20	1.55
0.1	0.99	+14	1.54
0.2	0.96	+8	1.51
0.3	0.91	+4.8	1.45
0.4	0.82	+2.7	1.37
0.5	0.707	+1.25	1.27
0.6	0.53	+0.35	1.15
0.707	0	—	1.0
1.0	0	—	0.644

Returning to our original problem whose response is shown in Figure 8-4, we see that M_p exceeds the +2 dB set as an upper limit for good response. The gear train or amplifier gain can be changed and the calculations repeated for each value of K_o to be analyzed. A new set of closed-loop data and a new closed-loop plot of A/θ is necessary for each value of K_o chosen until the desired response is obtained. In the next section, a graphical solution using the popular Nichols chart in combination with the Bode plot will provide a simple method for optimization of system gain to provide the desired closed-loop output response characteristic.

GLOSSARY

G: When dealing with a simple automatic control system, it represents the complex or frequency-dependent characteristics of a transfer function that affects the output amplitude and phase response as the frequency is changed.

H: Percent of the output fed back to the input. It compares to the β of a voltage amplifier with feedback. Sometimes referred to as the "feedback ratio."

K_o: The proportionality constant: the forward transfer characteristic of a system including each element and subsystem that makes up a complete system. It is modified by G, the frequency-dependent portion of the forward transfer characteristic.

M_p: The amplitude at ω_p expressed in decibels.

ω_B: The bandwidth of any open- or closed-loop system and is the frequency at which the output is 0.707 of the midfrequency response; the -3-dB down point of the amplitude response curve.

ω_p: The frequency of maximum amplitude under steady-state conditions defined as $\omega_p = \omega_n\sqrt{1 - 2\zeta^2}$.

ω_n: The resonant frequency $1/\sqrt{LC}$ of an electrical resonant system—the $\sqrt{K_o/\tau_m}$ of a closed-loop automatic control system where $A/\underline{\theta} = A/\underline{-90°}$.

ω/ω_n: A normalized ratio when $\omega/\omega_n = 1$ so that universal graphs can be prepared usuable at any resonant frequency (ω), by determining the normalizing constant n.

ζ: The damping factor, also the damping ratio. Defines the rate at which the output of a damped sine wave decreases per cycle; different from a, the damping coefficient, which defines the rate the output of a damped sine wave decreases with respect to time (t).

PROBLEMS

8.1 (a) Given the transfer function $140/[s(0.02s + 1)]$, determine the closed-loop transfer function with $H = 100\%$.

(b) What is the natural resonant frequency of the system under closed-loop conditions?

(c) What is the damping factor?

(d) What are ω_p, M_p, and ω_B? (Use calculations and verify answers by referring to Table 8-3.)

8.2 What happens to ω_p and M_p of Problem 8.1 when the gain is increased by a factor of $+17$ dB to 1000.

8.3 Given:

$$K \cdot G = \frac{40}{s(0.5s + 1)}$$

For position control operation, determine the bandwidth of the system using Table 8-3. (*Hint*: Find ω_n and ζ.) Verify your answer using equation (8-41).

8.4 Verify that ω_B/ω_n for $\zeta = 0.707$ is equal to 1.0.

8.5 Verify that ω_B/ω_n for $\zeta = 1$ is equal to 0.707.

8.6 Using the transfer function obtained in Problem 8.1, establish A/θ for a number of frequencies above and below ω_m. Make a rough plot of the closed loop amplitude and phase response and compare your curve to that of Figure 8-5. How does it compare (in terms of the damping factor)?

8.7 Given the motor time constant $\tau_m = 0.2$ s:

  (a) What is the system gain (K_o) needed to provide a $\zeta = 0.5$?
  (b) What happens to the M_p if K_o is increased by 10 dB?
  (c) Verify your conclusions mathematically using the closed-loop transfer function with new K_o and ζ.
  (d) Prove mathematically that the phase shift is very close to $-90°$ at ω_n.

8.8 Given

$$TF = \frac{80}{s(0.06s + 1)}$$

Determine

  (a) ω_n
  (b) ζ
  (c) ω_d
  (d) M_p at ω_p.
  (e) What happens to M_p if K_o is reduced by 10 dB?

9

Graphical Solution of Closed-Loop Systems

9-1 INTRODUCTION

In Chapter 6, we were shown the ease with which the transfer function of an open-loop system could be analyzed by the use of the Bode plot. Instead of a point-by-point analysis of $A\underline{/\theta}$, straight-line approximations permitted a rapid solution of otherwise slow computations of response. In a like manner, a graphical solution is available for quickly determining the *closed-loop response* of a transfer function.

The plot of the open-loop data (readily obtained from a Bode plot of $A\underline{/\theta}$ versus ω) is transferred to a very particular type of graph, called the *Nichols chart*, from which the *closed-loop data* can be read directly from a second set of coordinates. Because the Nichols chart is so readily adaptable to a second-order system of the type we have been discussing, it will be used to verify many of the conclusions arrived at by the mathematical derivations in Chapter 8.

9-2 THE NICHOLS CHART

The Nichols chart uses a rectangular coordinate grid system for presenting open loop data (A/θ) versus frequency (ω) with θ in degrees as the x axis and A in dB as the y axis. Superimposed on the rectangular grid is a contour graph of A/θ for a closed-loop system $KG/(1 + KG)$ which is a plot in the polar plane. Both plots are normalized for $\omega/\omega_n = 1$, so that for any value of ω, A/θ open loop agrees with A/θ closed loop.

Nichols charts appear in several basic forms, of which Figure 9.1(a) and (b) are typical. Note that the x and y axes are in degrees and dB, respectively. These are the A/θ coordinates for a number of frequencies taken from the open-loop Bode plot of a system. Superimposed on the x and y coordinates (A/θ open-loop data) is a second graph of Gain $KG/(1 + KG)$ and Angle $KG/(1 + KG)$, from which the closed-loop system data for the same set of frequencies are read off directly.

9-3 PROCEDURE FOR USING THE NICHOLS CHART

The Nichols chart provides us with a simple, accurate, and flexible solution of the closed-loop response for a servo system with 100% feedback. It is especially valuable in setting the K_o (forward loop gain) so that ω_p does have approximately a $+2$-dB peak response. The method for setting the K_o so that ω_p does not exceed $+2$ dB is a simple procedure once a Nichols chart has been prepared for the particular system under consideration.

By the simple technique using the x and y axes for presenting open-loop data with a curve of ω versus A/θ, it is quickly possible to extrapolate the closed-loop data of A/θ for the same frequencies. Take the closed-loop transfer function of equation (8-12), which we analyzed so thoroughly in Chapter 8, and set up the open-loop transfer function:

$$\text{TF} = \frac{100}{s(0.05s + 1)} \qquad (9\text{-}1)$$

Figure 9-2 is the Bode plot of equation (9-1). The phase and frequency become critical near the corner frequency in the closed-

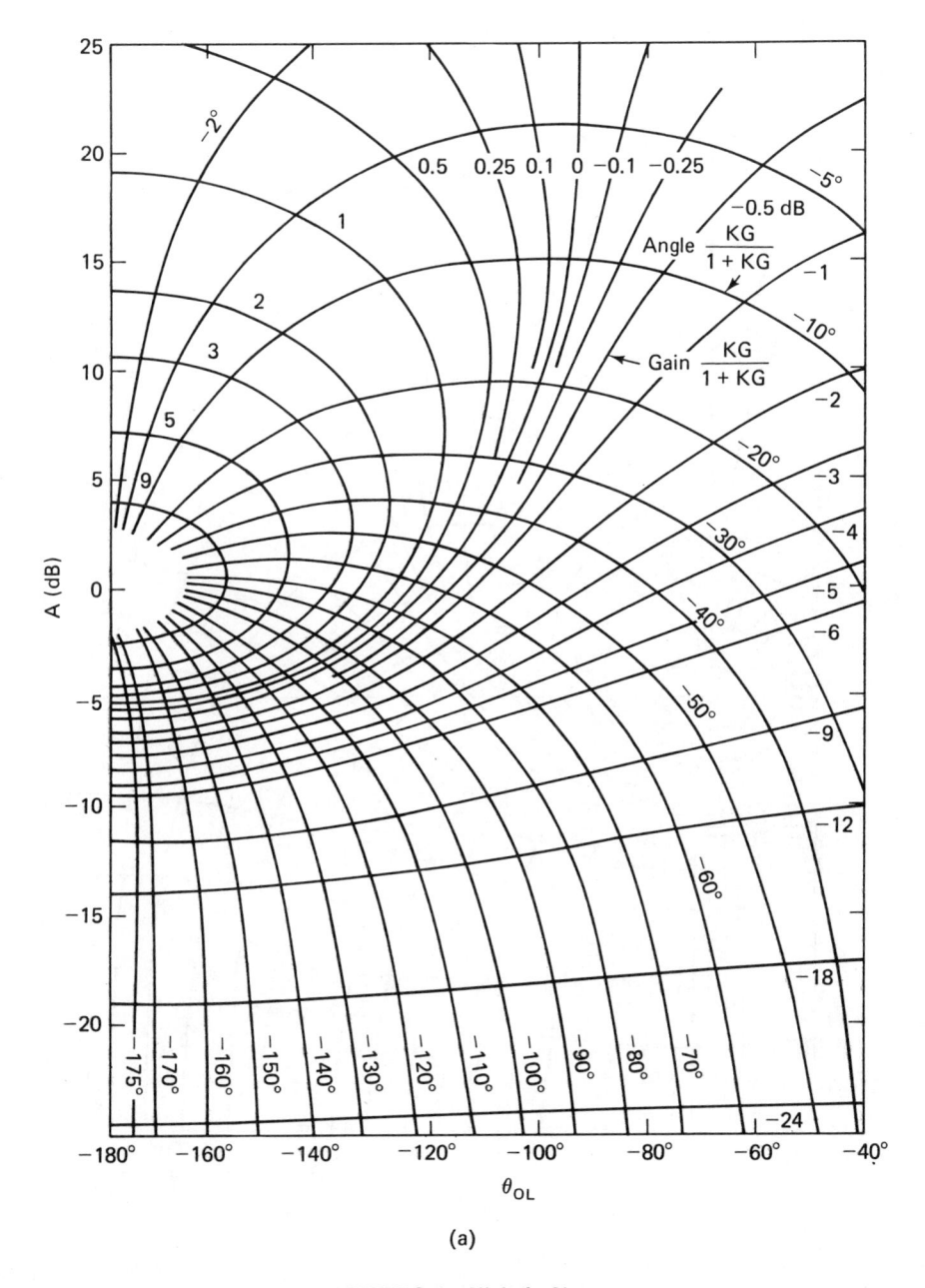

FIGURE 9-1 Nichols Chart.

(b)

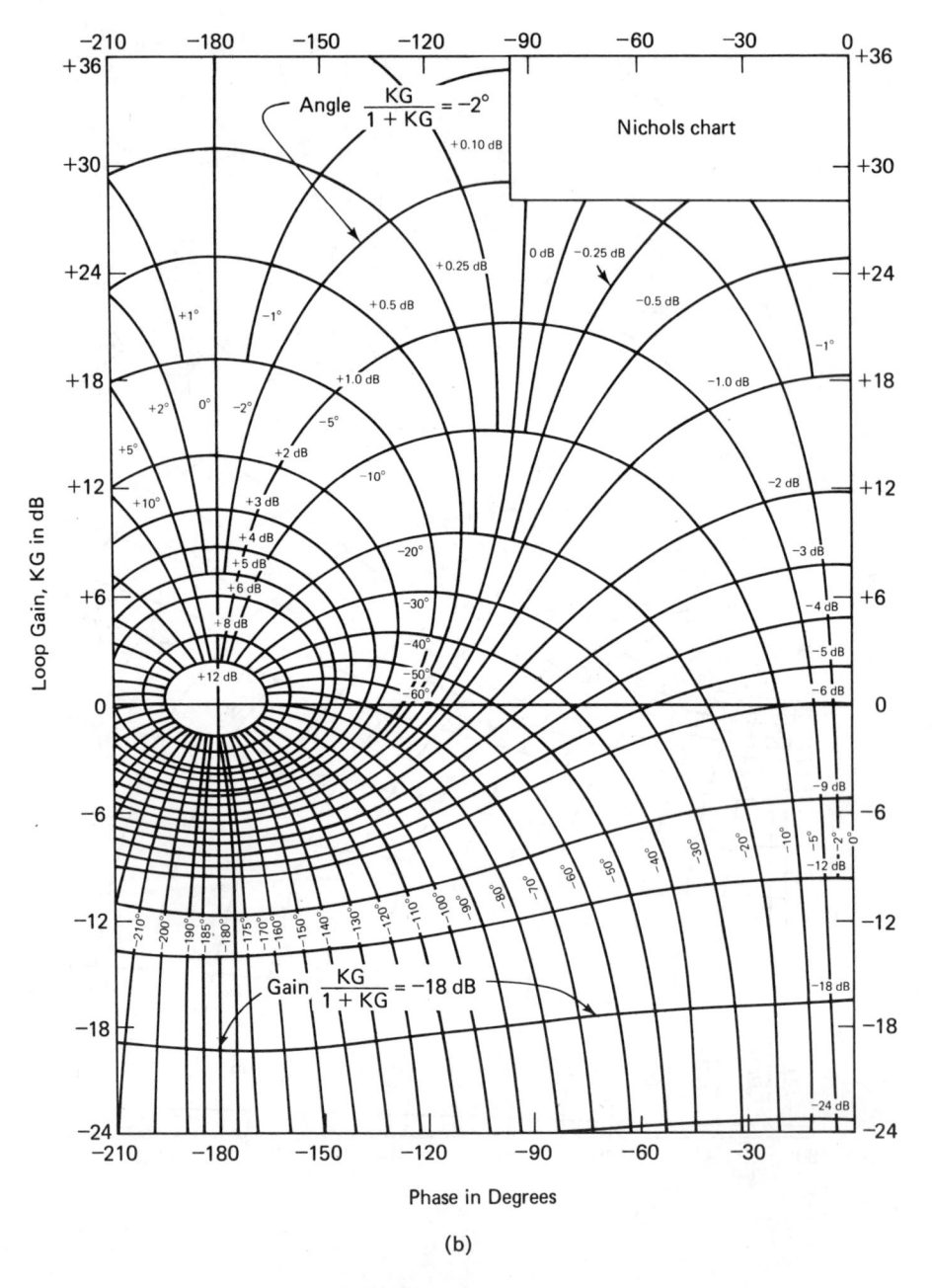

FIGURE 9-1 (Cont.)

Sec. 9-3   Procedure for Using the Nichols Chart   **167**

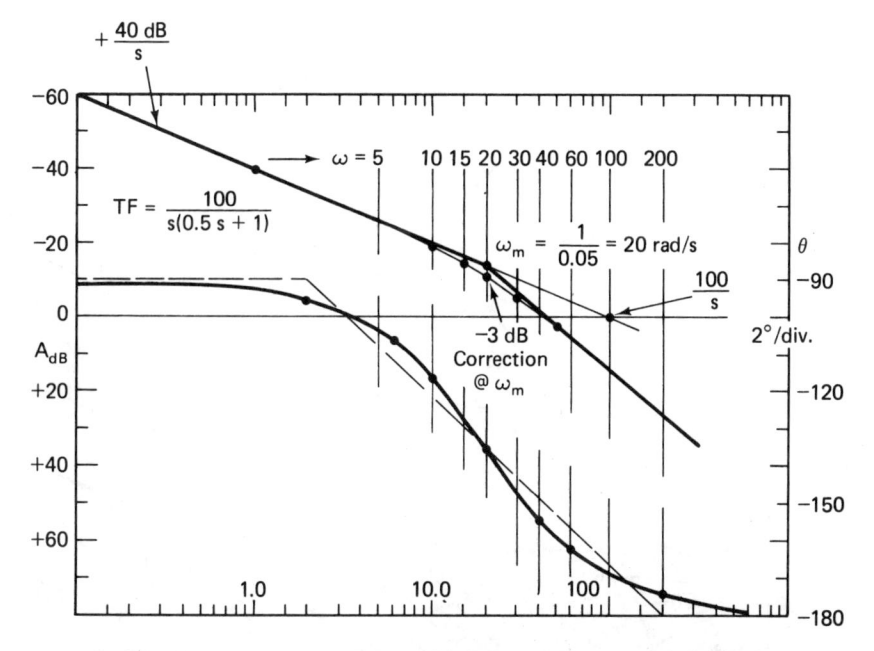

FIGURE 9-2 Bode plot of TF = $100/s(0.05s + 1)$ to produce Table 9-1.

loop analysis, and for this reason, it is necessary that the -3-dB correction be made at $\omega_m = 1/0.05$ or 20 rad/s, and that θ be corrected as per Table 3-3. From the plot of Figure 9-2, Table 9-1 is prepared.

TABLE 9-1 Open-loop data from Bode plot of Figure 9-2.

ω	A (dB)	θ_o (deg)
5	+26	-106
10	+19	-116
15	+14	-126
20	+11	-135
30	+5	-146
40	+1	-154
60	-5	-162
100	-14	-168
200	-26	-174

Transposing the open-loop data of Table 9-1 to the Nichols chart, Figure 9-3, provides the closed-loop data of A and θ for the

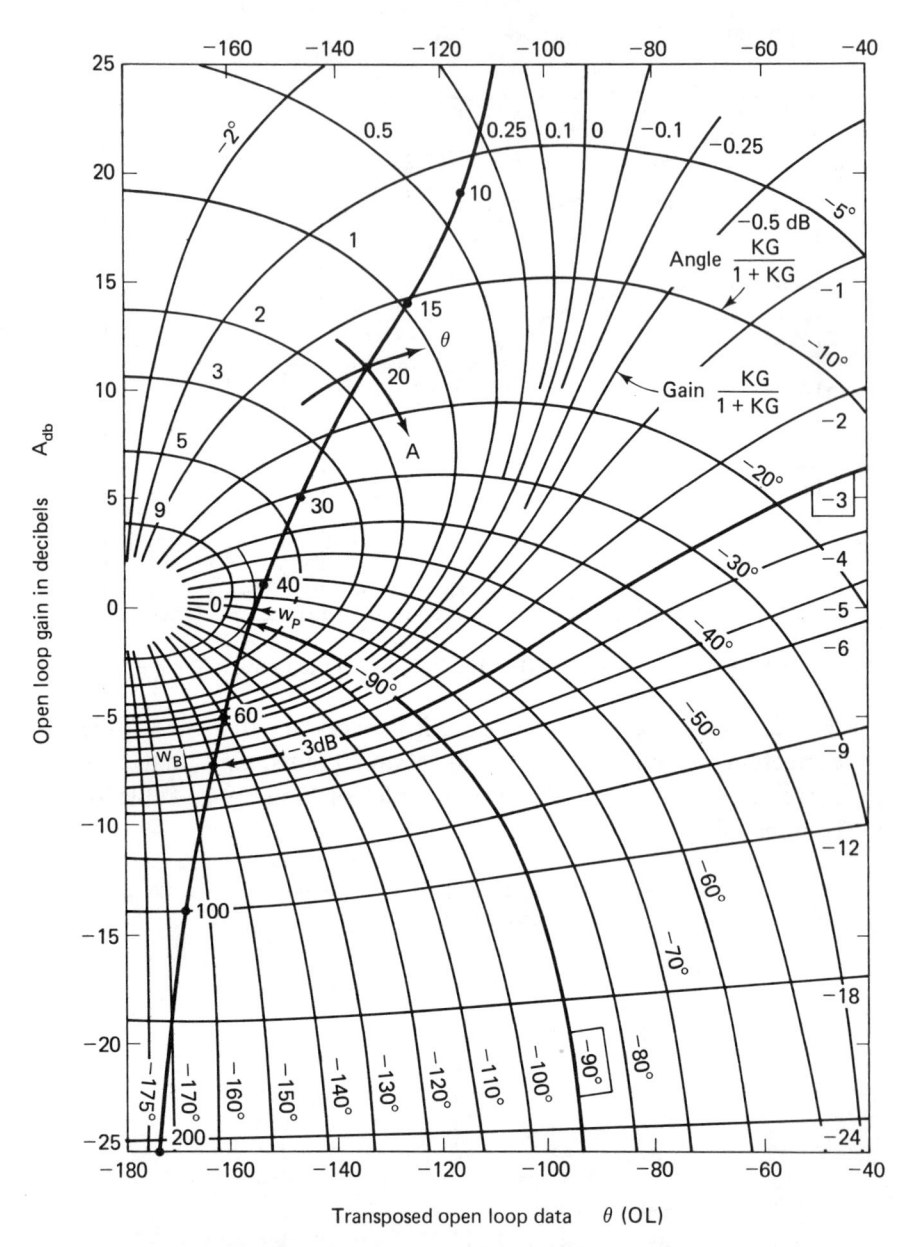

168 Graphical Solution of Closed-Loop Systems   Ch. 9

FIGURE 9-3 Transposed open-loop data of Table 9-2.

same set of frequencies that are shown in Table 9-2 (note heavy lines at $\omega = 20$ rad/s, A/θ (open loop) $= +11$ dB$/-135°$ and A/θ (closed loop) $= +1.6$ dB$/-15°$. Extending this procedure for all listed values of ω, we obtain the closed-loop data of Table 9-2. A comparison of these data to the *point-by-point* calculations of Table 8-2 will show they are almost identical (the only difference being reading errors from the Nichols chart).

TABLE 9-2 Closed-loop data from Figure 9-3.

for TF = $\dfrac{100}{s(0.05s + 1) + 100}$

ω (rad/s)	A (dB)	θ (deg)
10	+0.3	−6
15	+0.9	−10
20	+1.8	−16
30	+4	−30
($\omega_p = 43$)	+7.6	−80
60	+0.5	−140
100	−12	−165
200	−26	−174

9-4 INTERPRETING THE NICHOLS CHART

Before proceeding with the plot of Table 9-2, some important data can be taken directly from the Nichols chart of Figure 9-3. For example, $\omega_p \approx 44$ rad/s with $A/\theta \approx +7.5$ dB$/-80°$ (see tangent at point "0", since this is the point of maximum closed-loop amplitude). Also, it has been shown that at ω_n, $\theta = -90°$. ω_n can be found by determining where the $-90°$ closed-loop phase curve crosses our plot. (In this case $\omega_n \approx 45$ rad/s.) Finally, since ω_B is at -3 dB, from the Nichols chart we can find the closed-loop bandwidth by extending -3 dB [Gain $KG/(1 + KG)$], shown in heavy lines through our plot (see point $\omega_B \approx 68$ rad/s).

For a clearer presentation of the data above, a Bode plot presentation of the A/θ data (Figure 9-4) for the closed-loop data of Table 9-2 is needed. From it, ω_p, ω_n, ω_B and θ are much more readily recognized.

A quick mathematical check will show that the results obtained from the chart are indeed quite accurate. From the previous chapter, equations were established to determine ω_n, ζ, and ω_p.

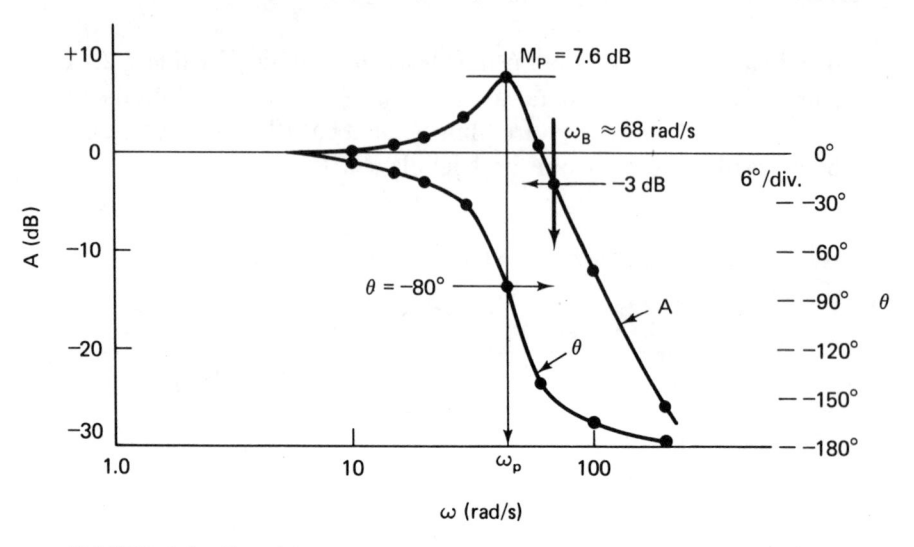

170 *Graphical Solution of Closed-Loop Systems* *Ch. 9*

FIGURE 9-4 Closed-loop data from Nichols chart for TF = $100/[s(0.05s + 1) + 100]$.

With $K_o = 100$ and $\tau_m = 0.05$ s:

$$\omega_n = \sqrt{\frac{K_o}{\tau_m}} = \sqrt{\frac{100}{0.05}} = 44.7 \text{ rad/s} \qquad (9\text{-}2)$$

$$\zeta = \tfrac{1}{2}\omega_n \tau_m = \frac{1}{(2)(44.7)(0.05)} = 0.224 \qquad (9\text{-}3)$$

from which

$$\omega_p = \omega_n\sqrt{1 - 2\zeta^2} = 44.7\sqrt{1 - 2(0.244)^2} = 42.4 \text{ rad/s} \qquad (9\text{-}4)$$

The amplitude M_p at ω_p is found using equation (8-20):

$$A\underline{/\theta} = \frac{K_o}{-\omega^2 \tau_m + j\omega + K_o}$$

$$= \frac{100}{-(42.4)^2(0.05) + j42.4 + 100} \qquad (9\text{-}5)$$

and

$$M_p = \frac{100}{j42.4 + 10.1} = +7.2 \text{ dB}\underline{/-76.6}$$

ω_B can be verified by using equation (8-37). With $\zeta = 0.224$,

$$\omega^4 - \omega^2(2 - 4\zeta^2) - 1 = 0 \qquad (9\text{-}6)$$

we obtain

$$\omega^4 - \omega^2[2 - 4(0.224)^2] - 1 = \omega^4 - 1.8\omega^2 - 1 = 0$$

$$a = 1$$
$$b = -1.8$$
$$c = -1$$

Substituting a, b, and c into the quadratic equation,

$$\omega^2 = \frac{-b \pm \sqrt{b^2 - 4ac}}{2a} = \frac{+1.8 \pm \sqrt{(1.8)^2 - 4(-1)}}{2} \quad (9\text{-}7)$$
$$= \frac{1.8 \pm \sqrt{7.24}}{2} = 2.24$$

and $\omega = 1.5$, with $\omega = \omega_B/\omega_n$, $\omega_B = \omega_n(1.5)$ and

$$\omega_B = 44.7(1.5) = 67 \text{ rad/s} \quad (9\text{-}8)$$

From the Nichols chart, the mathematical derivations and the closed-loop plot of Figure 9-4, we see that M_p is greater than the upper limit of $+2$ dB that we established earlier as the desirable upper limit of response under steady-state conditions. It should be apparent by now that a decrease in K_o is necessary to bring down M_p, the peak amplitude response.

9-4-1 Determining K_o from the Nichols Chart

The Nichols chart provides a simple, accurate, and flexible solution for the closed-loop response as the K_o or forward loop gain is changed. Recall that use of the Nichols chart is a technique or graphical solution, and therefore the procedure that follows is a mechanical method.

Let us take the system of equation (9-1) whose closed-loop response is shown by Figure 9-4 and decide to limit M_p to $+2$ dB. Basically, the plot of Figure 9-3 provides the information needed to complete the solution. Figure 9-5 is a duplicate of Figure 9-3 except for the emphasis of the $+2$ dB [Gain $KG/(1 + KG)$] circle (curve B). It is only necessary to drop curve A so that it is tangent to curve B to obtain the decrease in dB necessary to limit ω_p to

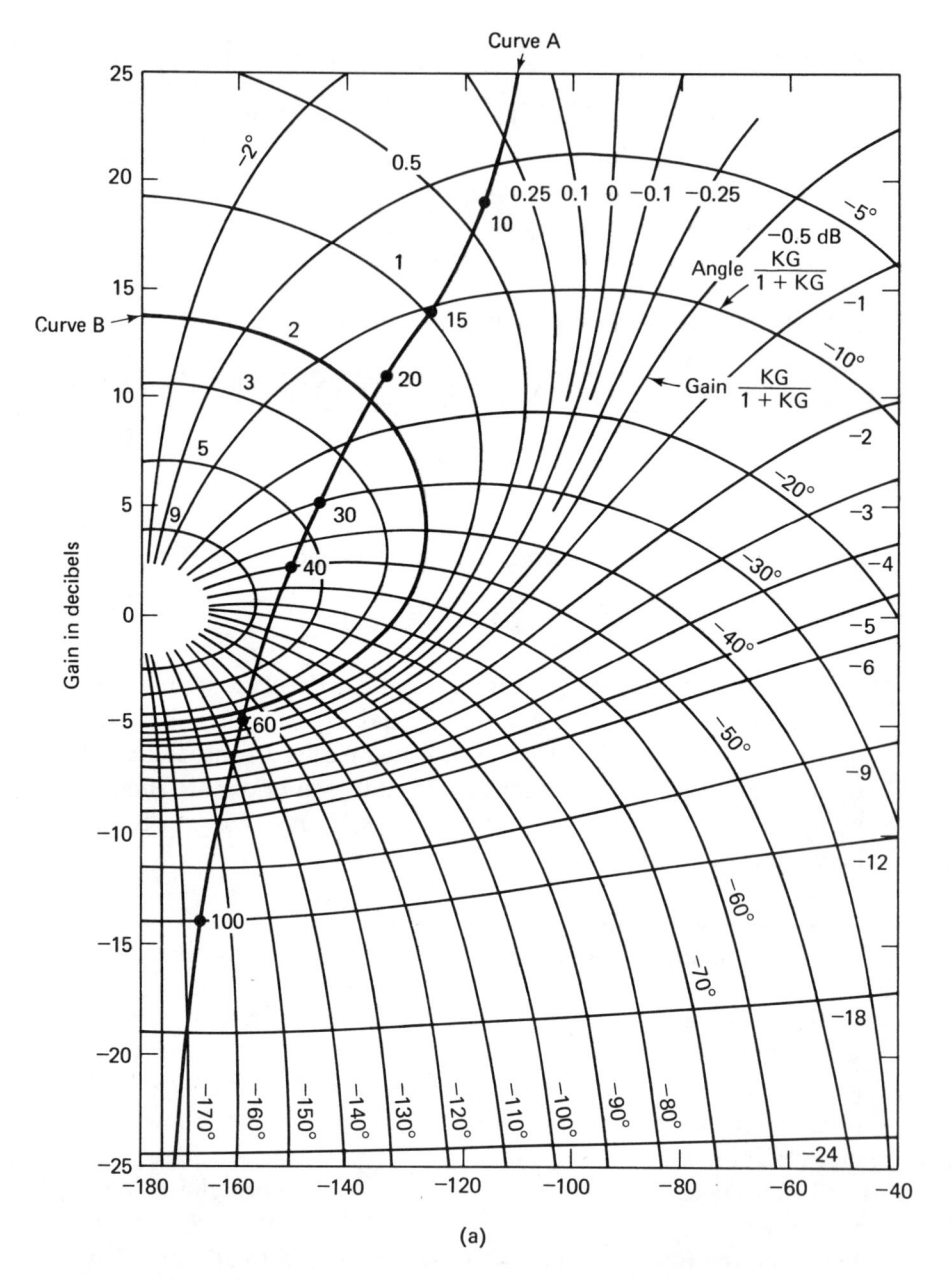

FIGURE 9-5 Determining K_o for M_p = +2dB with τ_m = 0.05s.

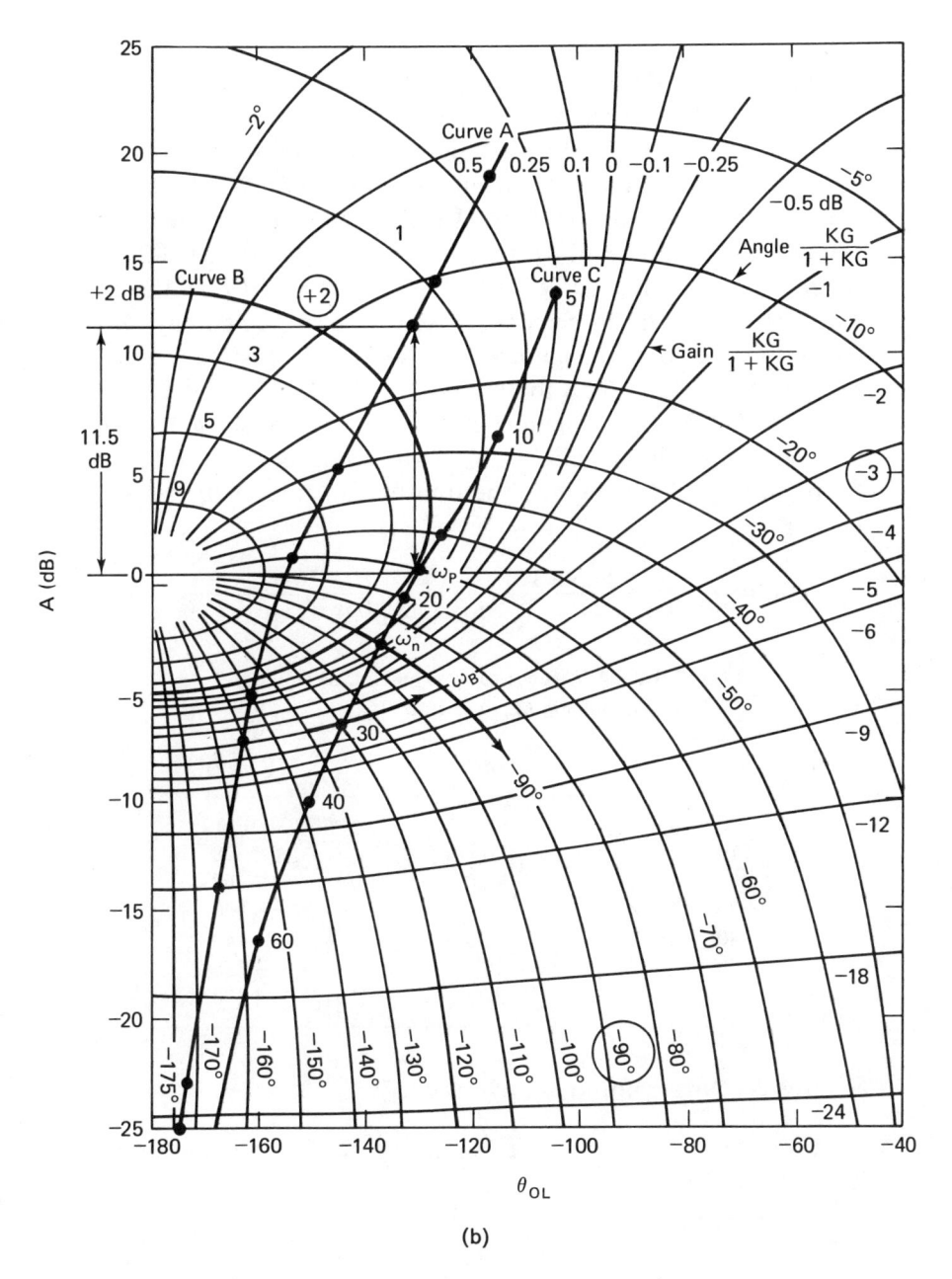

FIGURE 9-5 (Cont.)

+2 dB (see curve C). This can be fairly well approximated by observation. However, the standard procedure is to use a transparent sheet or a sheet of graph paper with the same x and y divisions as are used for the open-loop plot of ω versus $A/\underline{\theta}$ (see Figure 9-6). We again replot A and θ as in Figure 9-3. Now, by moving Figure 9-6 up or down, keeping the y axis in line, the amount moved up or down to obtain a tangent on the +2-dB closed-loop response (curve B) is the amount in dB that the K_o must be increased or decreased. Actual measurement by this method (and a look at Figure 9-5) shows the K_o must be reduced by approximately 11.5 dB to limit ω_p to +2 dB.

Table 9-3 is a compilation of the $A/\underline{\theta}$ data versus ω for the modified closed-loop gain of $K_o = 40 - 11.5 = 28.5$ dB taken from curve C of Figure 9-5. The transfer function of the revised system is

$$\text{TF} = \frac{+28.5 \text{ dB}}{s(0.05s + 1) + 28.5 \text{ dB}} = \frac{26.6}{s(0.05s + 1) + 26.6} \quad (9\text{-}9)$$

TABLE 9-3 $A/\underline{\theta}$ versus ω from curve C of Figure 9-5 limiting M_p to + 2 dB.

ω	A (dB)	θ (deg)
5	+0.2	−11
10	+0.90	−28
18	+2.0	−60
30	−3	−120
40	−7.5	−140
60	−15	−158
100	−24	−170

9-4-2 Determining ω_p, M_p, and ω_B

From the Nichols chart of Figure 9-5, the following information is obtained:

1. $\omega_p \approx 18$ rad/s.
2. $M_p \approx +2$ dB.
3. $\omega_B \approx 30$ rad/s.

To verify the accuracy of the data from the Nichols chart, for a

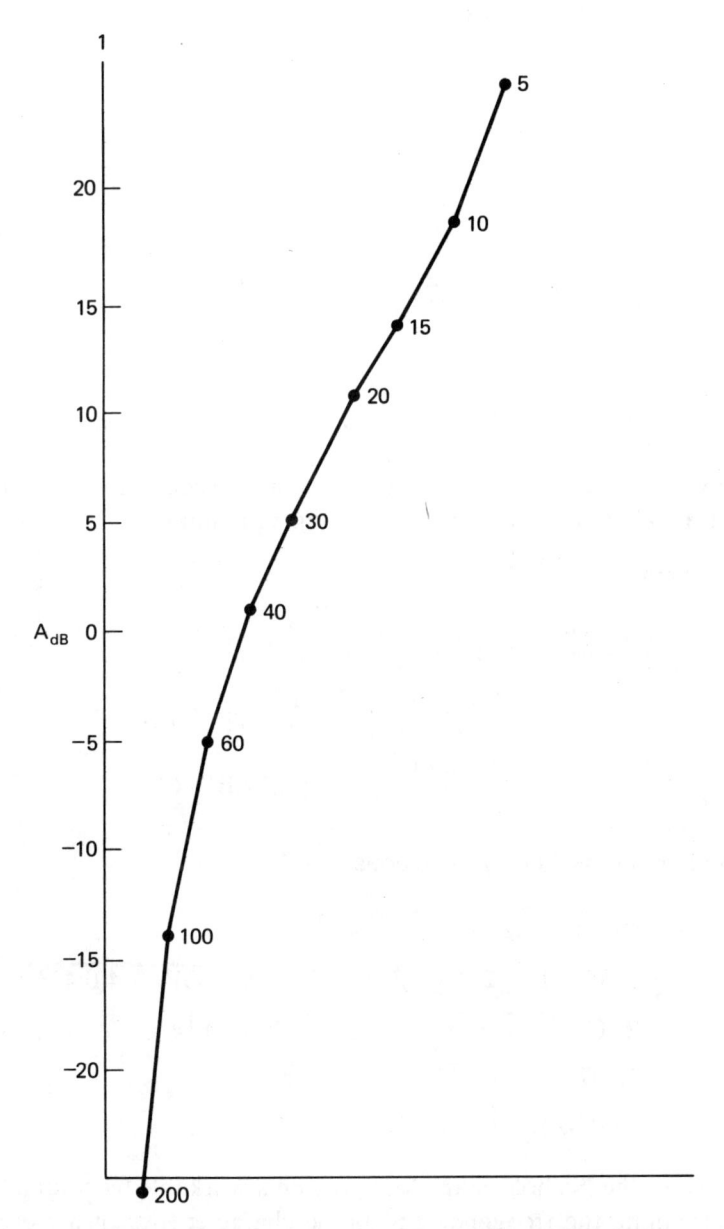

FIGURE 9-6 Amplitude and phase plot of TF = $100/s(0.05s + 1)$ used with Nichols chart to find K_o to limit M_p to +2dB.

$K_o = +28.5$ dB or 26.61 and a $\tau_m = 0.05$ s:

STEP 1

$$\omega_n = \sqrt{\frac{K_o}{\tau_m}} = \sqrt{\frac{26.61}{0.05}} = 23.07 \text{ rad/s}.$$

STEP 2

$$\zeta = \frac{1}{2\omega_n \tau_m} = \frac{1}{2 \times 23.07 \times 0.05} = 0.433$$

and

$$\omega_p = \omega_n\sqrt{1 - 2\zeta^2} = 23.07\sqrt{1 - 2(0.433)^2} = 18.24 \text{ rad/s}$$

Now given $\omega_p = \omega = 18.24$ rad/s, the amplitude ($M_p = +2$ dB) and the phase can be determined using equation (9-5):

STEP 3

$$A\underline{/\theta} = \frac{K_o}{-\omega^2 \tau_m + j\omega + K_o}$$

$$= \frac{26.61}{-(18.24)^2(0.05) + j18.24 + 26.61} \quad (9\text{-}10)$$

$$= \frac{26.61}{9.98 + j18.24} = +2.1 \text{ dB}\underline{/-61°}$$

To verify $\omega_B \approx 30$ rad/s, use equation (8-41):

$$\omega_B = \omega_n\sqrt{1 - 2Z^2 + \sqrt{2 - 4Z^2 + 4Z^4}}$$

$$= 23.07\sqrt{1 - 2(0.433)^2 + \sqrt{2 - 2(0.433)^2 + 4(0.433)^4}}$$

$$= 23.07\sqrt{1 - 0.375 + \sqrt{2 - 0.75 + 0.14}} \quad (9\text{-}11)$$

$$= 23.07\sqrt{1 - 0.375 + 1.179}$$

$$= 23.07\sqrt{1.343} = 30.99 \text{ rad/s}$$

Thus, the Nichols chart does provide a workable technique for determining the frequency and phase characteristics of a closed-loop system. There are several other popular techniques used to reach the same conclusions. The two most commonly used, in addition to the Nichols chart, are the Nyquist plots and the root-locus method of analysis. In an introductory course to automatic control systems, an understanding of one graphical technique for

GLOSSARY

Angle $[KG/(1 + KG)]$: The phase angle obtained when solving for A/θ of a closed-loop system with 100% feedback.

Gain $[KG/(1 + KG)]$: The forward gain characteristic A (in decibels) when solving for A/θ of a closed-loop system with 100% feedback, where $K \cdot G$ equals the open-loop gain characteristics.

Nichols Chart: A graph using a rectangular coordinate grid for the open-loop plot of A/θ at any frequency (ω) with θ as the x axis and A in dB as the y axis. Superimposed on the rectangular grid is a contour graph of A/θ for a closed loop system $[KG/(1 + KG)]$ which is a plot in the polar plane. Both plots are normalized for $\omega/\omega_n = 1$ so that for any value of ω, A/θ open loop agrees with A/θ closed loop.

PROBLEMS

9.1 Given:

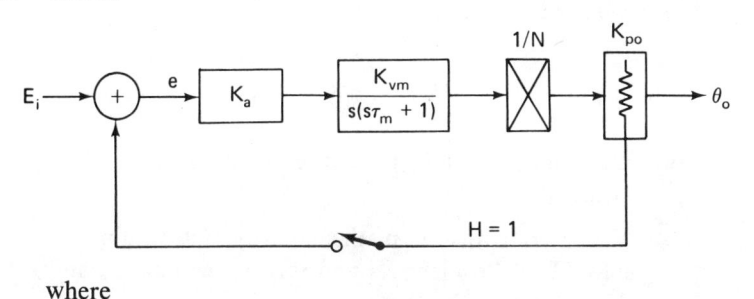

where

$$\text{TF} = \frac{\theta_o}{E_i} = K_a \cdot \frac{K_{vm}}{s(s\tau_m + 1)} \cdot \frac{1}{N} \cdot K_{po}$$

with

$$K_a = 25$$
$$K_{vm} = 15/[s(0.2s + 1)]$$
$$N = 50$$
$$K_{po} = 5\text{V/rad}$$

(a) Prepare the open-loop Bode plot of the transfer function (Correct both the amplitude and phase curves.)

(b) Calculate ω_n and ω_p.

(c) Prepare a table of $A/\underline{\theta}$ versus ω for several octaves above and below ω_m.

(d) Using a Nichols chart, prepare the closed-loop response by finding corresponding values of $A/\underline{\theta}$ at frequencies chosen in part (c).

(e) On the open-loop Bode plot of part (a), show the closed-loop response of the system.

(f) From the graph, determine ω_n, ω_p, M_p and ω_B.

(g) Verify M_p using ω_p found in part (b).

9.2 Using the Nichols chart from Problem 9.1,

(a) Adjust K_o so peak response is only $+2$ dB (See Figure 9.6.)

(b) Verify mathematically the new K_o will result in M_p of $+2$ dB. (first determine new ω_n and ζ).

9.3 Verify mathematically the bandwidth (closed-loop) of the system of Problem 9.1 using calculated ω_n and ζ and compare to results obtained in part (f) of Problem 9.1.

9.4 Repeat all the steps of 9.1 when

$$\text{TF}_{\text{OL}} = \frac{40}{s(0.1s + 1)}$$

9.5 (a) How much must K_o be reduced in Problem 9.4 for M_p to equal $+2$ dB?

(b) Verify your answer mathematically, substituting new K_o into TF of Problem 9.4 and solving for new ω_p and then M_p.

10

Transient Analysis

10-1 INTRODUCTION

In the analysis of closed-loop systems, two methods have been developed for determining the overall system response, a mathematical approach using complex algebra and the more convenient and popular Nichols chart translation of open-loop data to closed-loop data—with its flexibility for modifying system gain to provide a desired peak response.

A third very interesting approach to closed-loop analysis is the use of the step input, a suddenly applied force, torque, or more specifically a voltage, as shown in Figure 10-1, and the prediction of the system output response as a result of this form of excitation.

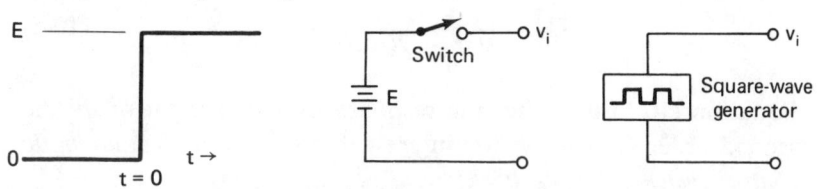

FIGURE 10-1 Step input of voltage E/S.

10-2 EFFECT OF A STEP INPUT ON SYSTEM RESPONSE

Several earlier conslusions will be restated to clarify the developments that follow. In Section 4-7, we spoke of the step input and the damping coefficient (*a*). This produced a damped sine-wave output of current described by

$$I(s) = \frac{E(s)}{Z(s)} = \frac{E}{L} \cdot \frac{1}{s^2 + R/L \cdot s + 1/LC} \tag{10-1}$$

The actual output we are interested in is that across the capacitor, which is equal to

$$v_o(s) = I(s) \cdot Z_o(s) \quad \text{where} \quad Z_o(s) = \frac{1}{sC} \tag{10-2}$$

Substituting equation (10-1) for $I(s)$ into equation (10-2), we obtain

$$v_o(s) = \frac{1}{sC} \cdot \frac{E}{L} \cdot \frac{1}{s^2 + R/Ls + 1/LC} \tag{10-3}$$

For $E = 1 = a$ unit step input,

$$TF = \frac{1}{LC} \cdot \frac{1}{s(s^2 + R/Ls + 1/LC)} \tag{10-4}$$

Substituting into equation (10-4) the following known relationships:

$$\frac{1}{LC} = \omega_n^2 = \text{natural resonant frequency}$$

$$\frac{R}{L} = \frac{1}{\tau} = 2\zeta\omega_n \quad \text{[see equation (8-18)]}$$

The transfer function becomes

$$TF = \frac{\omega_n^2}{s(s^2 + 2\zeta\omega_n \cdot s + \omega_n^2)} \tag{10-5}$$

Equation (10-5) describes the response to a step input of *all* the second-order systems we are interested in (*including the automatic control system with feedback*).

Sec. 10-2 Effect of a Step Input on System Response 181

The Laplace transform of equation (10-5)* provides

$$f(t) = \left(1 + \frac{1}{\sqrt{1-\zeta^2}}\right) e^{-\zeta\omega_n \cdot t} \sin(\omega_n \sqrt{1-\zeta^2} \cdot t - \psi) \quad (10\text{-}6)$$

where $\quad \psi = \tan^{-1}\left(\frac{\sqrt{1-\zeta^2}}{-\zeta^2}\right)$

$\quad\quad\quad \omega_d = \omega_n \sqrt{1-\zeta^2}$

This can be simplified to

$$v_o(t) = 1 + A \cdot e^{-\zeta\omega_n \cdot t} \sin(\omega_d \cdot t - \psi) \quad (10\text{-}7)$$

where $\quad \psi \text{ (rad)} = \tan^{-1} \frac{\sqrt{1-\zeta^2}}{-\zeta^2}$

$\quad\quad A$ = maximum peak amplitude of the sine wave
$\quad\quad\quad = 1$

Equation (10-7) is made up of three parts:

1. The constant $+1$:

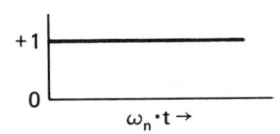

2. A sine wave ω_d decaying at the exponential rate of $e^{-\zeta\omega_n \cdot t}$

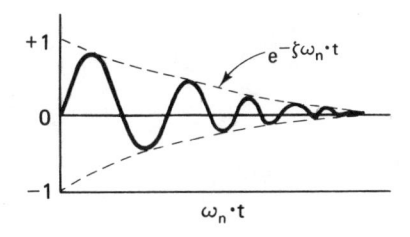

3. A phase delay $(-\psi = \tan^{-1} \sqrt{1-\zeta^2}/(-\zeta^2))$ in radians.

The three parts of equation (10-7) combine to produce the output response of a damped resonant system when energized by a

* From Floyd E. Nixon, *The Handbook of Laplace Transformation*, 2nd ed. (Prentice-Hall, Inc., Englewood Cliffs, N.J., 1965), derivation, p. 66, transform pair 00.101, p. 192.

182 *Transient Analysis* *Ch. 10*

step input. Figure 10-2 is a normalized plot of the output response for such a system with a specific damping factor of $\zeta = 0.2$. Referring to the figure:

1. The frequency of the damped wave (ω_d) in terms of ω_n and $\zeta = 0.2$ is

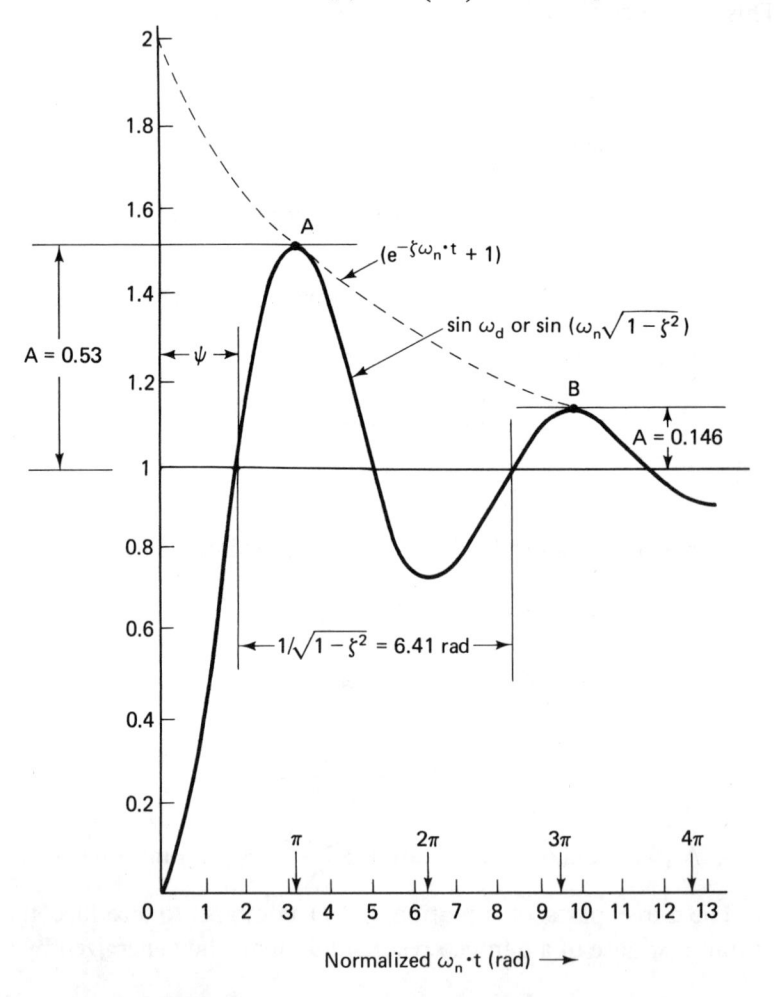

$$\omega_d = \omega_n\sqrt{1 - \zeta^2}$$
$$= \omega_n\sqrt{1 - (0.2)^2} = 0.98\omega_n \qquad (10\text{-}8)$$

FIGURE 10-2 Graphical presentation of Equation (10-6) $(1 + A \cdot e^{-\zeta\omega_n \cdot t} \sin \omega_d - \psi)$.

and

$$\frac{\omega_d}{\omega_n} = 0.98$$

As the damping increases, ω_d decreases until for $\zeta = 1$ (there is no frequency component present in the output), the system is considered to be *critically damped*.

2. The sine wave is delayed by ψ and for $\zeta = 0.2$,

$$\psi = \tan^{-1} \cdot \frac{\sqrt{1-\zeta^2}}{-\zeta^2} = \tan^{-1} \cdot \frac{\sqrt{1-(0.2)^2}}{-(0.2)^2} \quad (10\text{-}9)$$
$$= \tan^{-1}(-24.5) = 1.55 \text{ rad}$$

3. The rate of decay of the damped wave ω_d is defined by the boundary of the exponential $e^{-\zeta\omega_n \cdot t}$. With $\omega_d = \omega_n\sqrt{1-\zeta^2}$, for any value of ζ greater than zero, ω_d is a lower frequency than ω_n. The time of one cycle is inversely proportional to the frequency; therefore, one cycle of ω_d requires $1/\sqrt{1-\zeta^2}$ more time to complete than ω_n. Thus in terms of $\omega_n \cdot t$, where

$$\omega_n \cdot t = \frac{\text{rad}}{\text{s}} \cdot \text{s} = \text{rad} \quad (10\text{-}10)$$

the peaks of the damped wave of Figure 10-2 are

$$\text{first peak at } \omega_n \cdot t = \frac{\pi}{\sqrt{1-\zeta^2}} \text{ rad}$$

$$\text{second peak at } \omega_n \cdot t = \frac{3\pi}{\sqrt{1-\zeta^2}} \text{ rad}$$

and subsequent peaks every 360° or 2π rad, until they disappear.

The amplitude (A) of the first peak of a damped wave with a step input applied for any value of ζ is

$$A = 1 + e^{-\zeta\omega_n \cdot t} \quad (10\text{-}11)$$

where $\omega_n \cdot t$ for the first peak is $\pi/\sqrt{1-\zeta^2}$.

EXAMPLE 10-1

For a system with $\zeta = 0.2$, determine the amplitude of the first peak:

SOLUTION:

$$1 + e^{-0.2\pi/\sqrt{1-(0.2)^2}}$$
$$1 + e^{-0.2(3.142)/\sqrt{0.96}}$$
$$1 + e^{-0.644} = 1.53$$

This is shown as point A in Figure 6-2.

For the second peak with $\zeta = 0.2$ and $\omega_n \cdot t = 3\pi/\sqrt{1-\zeta^2}$,

$$1 + e^{-0.2(3\pi)/\sqrt{1-(0.2)^2}}$$
$$1 + e^{-1.86/0.98}$$

and $A = 1.146$, shown as point B in Figure 6-2.

In an automatic control system with feedback, which is also a second-order system (one capable of sustained oscillation), a step input produces identical results in the form of the output response. In practice, when a step input is applied as the command signal to a servo system, it would be most desirable for the output to respond as closely as possible to this command. However, the mass of the rotating components, with their inherent inertia, the limited power of the servo motor, and phase shift through various components and networks, combine to modify the output response. Ideally, the output from the follow-up potentiometer K_{po} should look like Figure 10-3(a). Actually, it varies, from the oscillating response of Figure 10-3(b) to the slow response of the overdamped output of Figure 10-3(c).

10-2-1 Response Time and Overshoot

Response time is the time it takes for an output to go from 10% to 90% of its final value when the input is a step. (It is different from the time constant, τ.) Since response time is often used in servo analysis, a look at Figure 10-4 may help clarify its meaning. Referring to curve (a) of Figure 10-4, response time is equal to

$$t_{r_1} = t_2 - t_1 \quad \text{seconds} \quad (10\text{-}12)$$

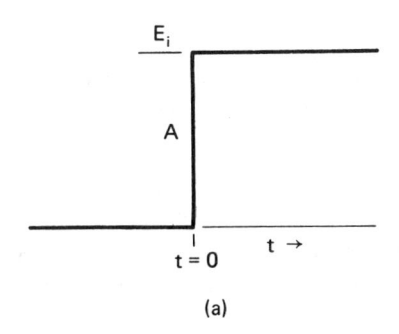

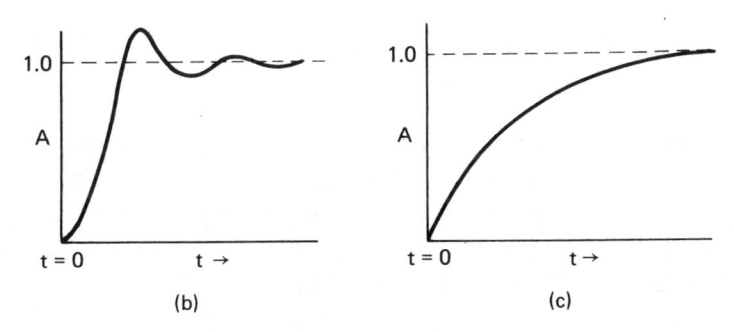

FIGURE 10-3 Transient response of a second order system to a step input: (a) input; (b) underdamped output; (c) overdamped outputs.

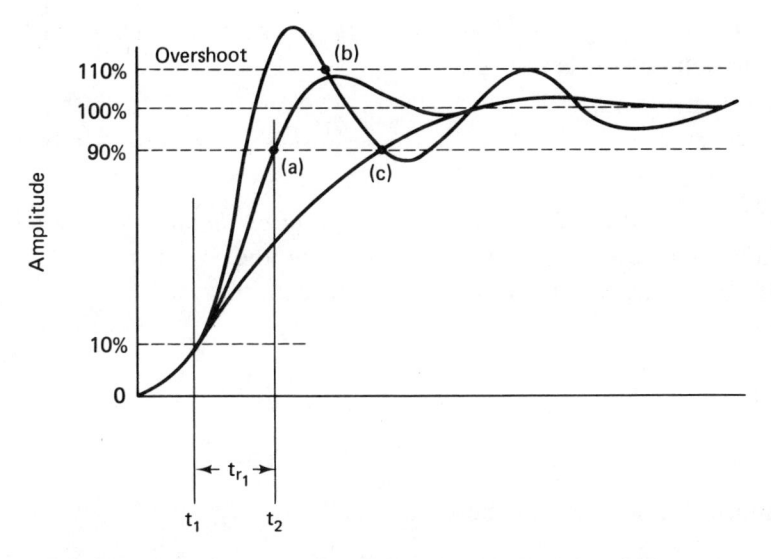

FIGURE 10-4 Response time: (a) fast response; (b) underdamped; (c) overdamped.

185

Again referring to Figure 10-4, *overshoot* is the percentage ratio between the amount the output exceeds its final steady-state value divided by the final steady-state value with a step input as the command signal. In the case of curve (a) of Figure 10-4, it is approximately 10%. A compromise must be made between speed of response, or the response time, and the amount of overshoot that can be tolerated in the output of an automatic control system.

10-2-2 Settling Time

Finally, once the damped wave ω_d is generated, it is important to know how many oscillations occur before the output is reduced to zero (taken as under 2%, although many designers use 5% as the stabilized condition). A look at the decaying exponential function of Figure 1-4 and Table 1-1 shows it takes three time constants for the output to reduce to less than 5% and four time constants for the output to decrease below 2%. Since the time of one cycle of oscillation is $1/f$, in terms of $\omega_d = 2\pi f$:

$$t_d = \frac{1}{f} = \frac{2\pi}{\omega_d} \tag{10-13}$$

In terms of the natural frequency (ω_n), where $\omega_d = \omega_n\sqrt{1-\zeta^2}$, equation (10-13) becomes

$$t_d = \frac{1}{f} = \frac{2\pi}{\omega_n\sqrt{1-\zeta^2}} \tag{10-14}$$

(where t_d is also referred to as the *period of oscillation*). We already know the rate of decay is described by the exponential $e^{-\zeta\omega_n \cdot t}$. If we assume that the output settles to under 2% when

$$e^{-\zeta\omega_n \cdot t} = e^{-4} \tag{10-15}$$

Then

$$\zeta\omega_n \cdot t = 4$$

and t_s, the settling time becomes

$$t_s = \frac{4}{\zeta\omega_n} \text{ seconds} \tag{10-16}$$

Thus, the number of cycles (f_s) required to reduce the output of a damped wave to less than 2% is equal to the total time divided by the time of one cycle:

$$f_s = \frac{4}{\zeta \omega_n} \div \frac{2\pi}{\omega_d}$$

or

$$f_s = \frac{4}{\zeta \omega_n} \cdot \frac{\omega_n \sqrt{1 - \zeta^2}}{2\pi} \tag{10-17}$$

and

$$f_s = \frac{2\sqrt{1 - \zeta^2}}{\pi \zeta} \tag{10-18}$$

In terms of a 5% lower limit, it takes three time constants and equation (10-17) becomes

$$f_s = \frac{3}{\zeta \omega_n} \cdot \frac{\omega_n \sqrt{1 - \zeta^2}}{2\pi} \tag{10-19}$$

making

$$f_s = \frac{1.5\sqrt{1 - \zeta^2}}{\pi \cdot \zeta} \tag{10-20}$$

Table 10-1 at the end of this section, lists the number of cycles needed to reduce the output to under 5% as ζ is varied.

EXAMPLE 10-2

(a) How many cycles does it take for a damped wave to decrease to under 2% of its maximum peak value when $\zeta = 0.4$?

(b) How many cycles does it take to decrease the damped output to under 5% of its maximum peak value when $\zeta = 0.4$?

SOLUTION:

(a) $f_s = \dfrac{2\sqrt{1 - (0.4)^2}}{(3.142)(0.4)} = 1.46$ cycles

(b) $f_s = \dfrac{1.5\sqrt{1 - (0.4)^2}}{(3.142)(0.4)} = 1.1$ cycles

10-2-3 Damped Response to a Unit Step Input

By definition, the first peak of Figure 10-2 is the first overshoot in percent. Several examples will provide the % overshoot as ζ is varied. Table 10-1 summarizes the measured and calculated overshoot versus ζ, and Figure 10-5 is a composite of the damped trains as ζ is varied. Two limits are present:

1. For $\zeta = 0$, $\omega_d = \omega_n\sqrt{1 - \zeta^2} = \omega_n$ and $A = 1$ or 100% overshoot is present.

2. For $\zeta = 1$, $\omega_d = \omega_n\sqrt{1 - 1} = 0$ and $A = 0$ and $\omega_d = 0$.

Since no oscillations are present, no overshoot can exist. For values of ζ other than 0 or 1, equation (10-11) provides the % overshoot.

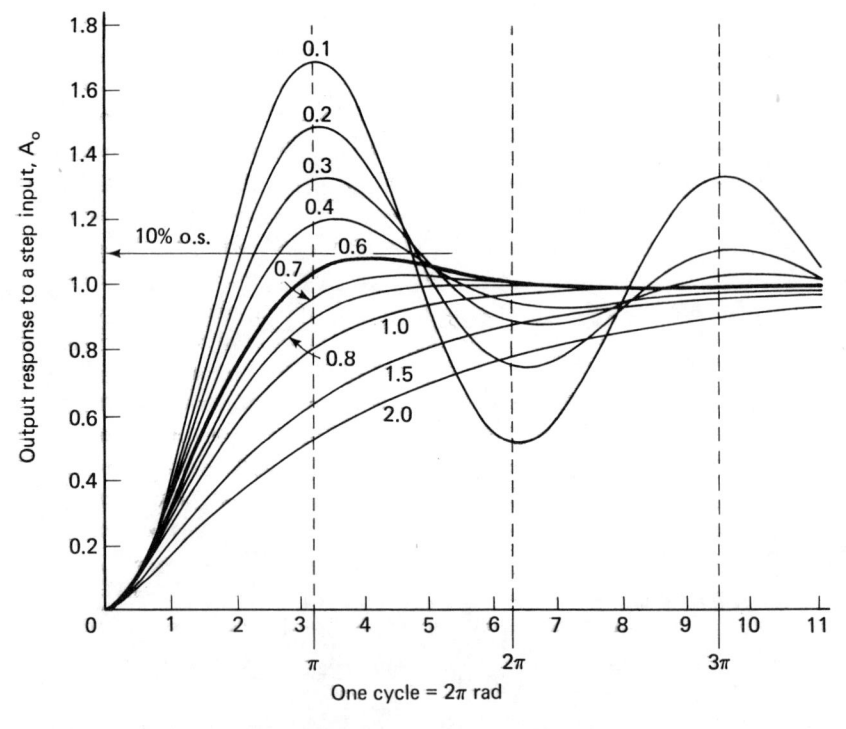

FIGURE 10-5 Plot of $A = 1 + e^{-\zeta\omega_n \cdot t}/\sqrt{1 - \zeta^2}$ for values of ζ (0.1 to 2.0).

Sec. 10-2   *Effect of a Step Input on System Response*   **189**

EXAMPLE 10-3

For $\zeta = 0.1$, how much overshoot is obtained?

$$A = e^{-\zeta\pi/\sqrt{1-\zeta^2}}$$
$$= e^{-0.1\pi/\sqrt{1-(0.1)^2}}$$
$$= e^{-0.1(3.142)/\sqrt{0.99}}$$
$$= e^{-0.316}$$
$$= 0.73 \quad \text{or} \quad 73\% \text{ overshoot}$$

EXAMPLE 10-4

For $\zeta = 0.2$,

$$A = e^{-0.2\pi/\sqrt{1-(0.2)^2}}$$
$$= e^{-0.64}$$
$$= 0.53 \quad \text{or} \quad 53\% \text{ overshoot}$$

EXAMPLE 10-5

For $\zeta = 0.5$,

$$A = e^{-0.5\pi/\sqrt{1-(0.5)^2}}$$
$$= 0.163 \quad \text{or} \quad 16.3\% \text{ overshoot}$$

This procedure is continued for solving the amount of overshoot for values of $\zeta < 1$, since there is no overshoot beyond $\zeta = 1$.

EXAMPLE 10-6

Using this same method, if the second overshoot occurs at 3π rad, what value of ζ would limit the second overshoot to under 2% so that actually only one overshoot is present?

SOLUTION:

$$A = e^{-\zeta(3\pi)/\sqrt{1-\zeta^2}}$$

or

$$0.02 = e^{-x} \quad \text{where } x = \zeta \cdot (3\pi/\sqrt{1-\zeta^2})$$

or

$$e^x = 50 \text{ (taking the log of both sides)}$$
$$x = \log e\,(50) = 3.912$$

Replacing x with $\zeta \cdot (3\pi/\sqrt{1-\zeta^2})$, we obtain

$$\zeta \frac{3\pi}{\sqrt{1-\zeta^2}} = 3.912$$

$$\zeta = \frac{3.912}{3} \cdot \sqrt{1-\zeta^2}$$

Squaring both sides yields

$$\zeta^2 = 0.17(1-\zeta^2) \quad \text{or} \quad \zeta^2 + 0.17\zeta^2 = 0.17$$

and

$$\zeta = \sqrt{\frac{0.17}{1.17}} = 0.4$$

EXAMPLE 10-7

As a matter of interest, what is the amplitude of the second overshoot for $\zeta = 0.5$?

SOLUTION:

$$A = e^{-\zeta(3\pi/\sqrt{1-Z^2})} = e^{-0.5[3\pi/\sqrt{1-(0.5)^2}]}$$
$$= 0.2\%$$

Table 10-1 and the graph of Figure 10-5 summarize the conclusions reached on the effects of ζ on system response. If the data of Table 10-1 were extended to include the overshoot at 3π rad, it develops into the universal plot of Figure 10-5. A study of Figure 10-5 indicates that for a fast response time, a ζ somewhere between 0.4 (which oscillates for a cycle or so) and 0.7 (which does not oscillate) is a good compromise.*

---

   * In a lathe cutting operation, no overshoot is tolerable and for a fast response time, a different design approach is used involving multiple closed loops in the system.

TABLE 10-1 Summation of closed loop system responses to a step input as a function of the damping factor ζ.

ζ	ω_d/ω_n	OVERSHOOT +dB	OVERSHOOT %	NUMBER OF CYCLES TO REDUCE OVERSHOOT 5%	NUMBER OF CYCLES TO REDUCE OVERSHOOT 2%
0.0	1	6.0	100	—	—
0.1	1	4.76	73	4.77	6.33
0.2	0.98	3.69	53	2.34	3.12
0.3	0.95	2.73	37	1.52	2.02
0.4	0.92	1.94	25	1.09	1.46
0.5	0.866	1.29	16	0.83	1.10
0.6	0.8	0.79	9.5	0.64	0.85
0.8	0.6	0.13	1.5	0.36	0.48
1.0	0.0	—	0	0.0	0.0

With $\zeta = 0.45$, the overshoot with a step input applied is only 20% and under steady-state conditions (the input is a sine wave), we know that a resonant peak (ω_p) occurs as indicated in Figure 8-5. This peak reaches a maximum amplitude (M_p) of under 2 dB with $\zeta = 0.45$.

EXAMPLE 10-8

(a) Determine the overshoot in a closed-loop system to a step input with $\zeta = 0.45$.

(b) What is ω_p in terms of ω_n for this system?

(c) What is the amplitude and phase ($A/\underline{\theta}$) at ω_p?

SOLUTION:

(a) A look at Table 10-1 shows that the overshoot for $\zeta = 0.45$ is close to 20% or $+1.6$ dB.

(b) $\omega_p = \omega_n \sqrt{1 - 2\zeta^2} = \omega_n \sqrt{1 - 2(0.45)^2} = 0.893\omega_n$

(c) To find M_p and the phase shift at ω_p, solve for $A/\underline{\theta}$ at ω_p. Use equation (8-27):

$$A/\underline{\theta} = \frac{1}{1 + j2\zeta\omega_n - \omega_n^2}$$

$$= \frac{1}{1 + j2(0.45)(0.893) - (0.893)^2} = +1.63 \text{ dB}/\underline{-75.9°}$$

10-3 TYPES OF RESONANT SYSTEMS

All oscillating systems, whether they be the tuned LC circuit, the spring–mass mechanical combinations, or a servo system with feedback, have the identical mathematical solution for their behavior toward a step input. This is because each of the three systems has its identical mathematical counterpart. Three of the four systems under consideration can be depicted as shown in Figure 10-6. (System III is of special interest because it contains most of the constants present in a servo system.) The components that make up each of the three systems shown in Figure 10-6 are summarized in Table 10-2.

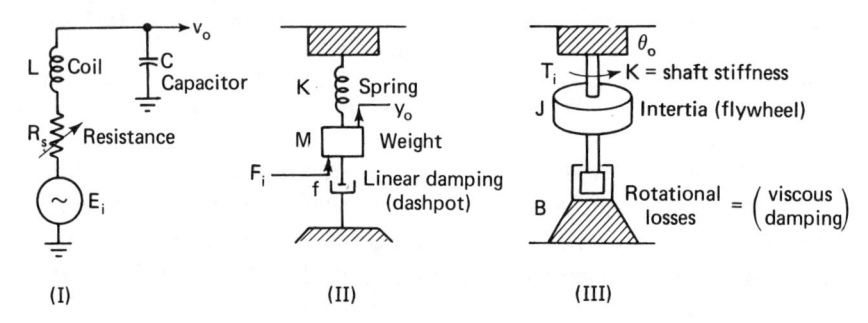

FIGURE 10-6 Three basic oscillating systems: (I) *LRC;* (II) spring—mass (linear); (III) spring-mass (rotational).

TABLE 10-2 Component groups that make up the three basic oscillatory systems.

System I
Coil $= L =$ inductance
Capacitor $= C =$ capacitance
Resistor $= R =$ resistance

System II
Weight $= M =$ mass
Spring $= K =$ stiffness
Dashpot $= f =$ mechanical resistance

System III
Flywheel $= J =$ inertia
Shaft $= K =$ torsional stiffness
Viscous damping $= B =$ rotational losses

Each system after being excited has stored energy: the capacitor stores energy (joules), a compressed spring stores energy depending on the force applied (newtons/meter or pounds/inch), and the flywheel can do work because it has inertia (newton-meters^2 or pound-inches^2). The mathematical units involved in each of the systems is listed in Table 10-3. Note in Table 10-3 that the analog for the capacitor (C) in systems II and III is $1/K$. It is customary to refer to the spring constant as the amount it compresses per unit weight (meters/newton or inches/pound). The output *after* the spring is released is newtons/meter or pounds of force per inch of compression, and therefore the energy released is the reciprocal of the input, or $1/K$.

TABLE 10-3 Standard units for the elements that make up the three basic oscillatory systems.

System I
L = henries
C = farads
R = ohms

System II
M = kilograms
$1/K$ = newton/meter
f = newton-seconds/meter

System III
J = newton-meter^2
$1/K$ = newton-meters/radian
B = newton-meters/radian/second

10-4 TRANSIENT ANALYSIS USING THE LAPLACE TRANSFORMS

We now have all the information needed to complete our study of a basic automatic control system under dynamic conditions. Basically, each of the three systems under analysis has its identical mathematical counterpart. Table 10-4 shows the common interrelationships that exist among the systems.

194   *Transient Analysis*   *Ch. 10*

TABLE 10-4   Summation of system parameters.

Type of System	Input	System Elements	Output
Electrical	E (volts)	L, C, R	$I =$ amperes
Spring–mass	F (force)	$M, 1/K, f$	$Y =$ meters
Rotary spring–mass	T (torque)	$J, 1/K, B$	$\theta =$ radians

Once an equation is established in terms of the natural resonant frequency ω_n and the damping factor ζ, each of the three systems can be evaluated in terms of the elements that make up the system. Let us begin with the electrical resonant system.

10-4-1   The Electrical Resonant System

In the beginning of this chapter, it was shown that equation (10-4) was equal to equation (10-5), from which (for any given electrical resonant system):

$$s^2 + \frac{R}{L} \cdot s + \frac{1}{LC} = s^2 + 2\zeta\omega_n \cdot s + \omega_n^2 \qquad (10\text{-}21)$$

It should be apparent from equation (10-21) that for the particular circuit, under consideration:

$$\frac{R}{L} = 2\zeta\omega_n \quad \text{and} \quad \omega_n = \frac{1}{\sqrt{LC}} \qquad (10\text{-}22)$$

making

$$\zeta = \frac{R}{(2\omega_n \cdot L)} \quad \text{and for } Q = \omega_n \cdot \frac{L}{R}, \zeta = \frac{1}{2Q} \qquad (10\text{-}23)$$

Since $R/L = 1/\tau$, then $1/\tau = 2\zeta\omega_n$, from which we obtain

$$\zeta = \frac{1}{2\omega_n \cdot \tau} \qquad (10\text{-}24)$$

Knowing ζ and ω_n, it is a simple matter to find ω_d, since

$$\omega_d = \omega_n\sqrt{1 - \zeta^2} \quad \text{(nots when } \zeta = 1, \omega_d = 0\text{)} \qquad (10\text{-}25)$$

In an electrical resonant system, *critical damping* ($\omega_d = 0$) occurs

Sec. 10-4   Transient Analysis Using the Laplace Transforms   195

when $\zeta = 1/2Q = 1$ or $Q = 0.5$ and no frequency component will be present in the output when a step input of voltage is applied to the circuit.

Using the same technique, we can now generate the equations for describing the response of the spring–mass system and the basic servo system to a command input signal.

10-4-2 The Spring–Mass System

In the following development, all units are in the meter-kilogram-second system.

1. When a spring is compressed by a force (F) through a distance (Y), it is very much like a charged capacity (C) which has potential energy in joules $(\frac{1}{2}CV^2)$, where 1 joule $=$ 0.736 foot-pound. The spring stiffness (K) is simply the compression of the spring in newtons/meter (or inches/pound) and the spring in compression contains the potential energy.

2. The mass (M) is displaced when the spring is released, and like the inductance (L), does not absorb energy but develops inertia, which stretches and then compresses the spring again and again. The mass would rise and fall through a distance (Y) almost forever except for minor losses in the mechanical system due to windage and molecular friction. By adding additional damping resistance (f) through a dashpot (like a shock absorber absorbing energy), the number of oscillations can be controlled as (f) is varied (the same as R in an electrical circuit).

Replacing the electrical units in equation (10-21) by its equivalent analog from Table 10-4, we obtain the second-order equation for the oscillating spring–mass resonant system, resulting in

$$Y(s) = \frac{K/M}{s(s^2 + (f/M)s + K/M)} = \frac{\omega_n^2}{s(s^2 + 2\zeta\omega_n \cdot s + \omega_n^2)} \quad (10\text{-}26)$$

from which

$$s^2 + \frac{f}{M} \cdot s + \frac{K}{M} = s^2 + 2\zeta\omega_n \cdot s + \omega_n^2 \quad (10\text{-}27)$$

so that

$$\omega_n^2 = \frac{K}{M} \quad \text{and} \quad \omega_n = \sqrt{\frac{K}{M}} \qquad (10\text{-}28)$$

and

$$2\zeta\omega_n = \frac{f}{M} \quad \text{making} \quad \zeta = \frac{f}{2\omega_n \cdot M} \qquad (10\text{-}29)$$

with

$$\frac{M}{f} = \tau, \text{ the time constant} \qquad (10\text{-}30)$$

Substituting $\omega_n = \sqrt{K/M}$ into equation (10-29) yields

$$\zeta = \frac{f}{2\sqrt{K/M}(M)} = \frac{f}{\sqrt{4KM}} \qquad (10\text{-}31)$$

For critical damping

$$\zeta = 1 = f/\sqrt{4KM} \quad \text{or} \quad f = \sqrt{4KM} \qquad (10\text{-}32)$$

10-4-3 The Rotating System

As we did in Section 10-4-2 if we replace the mass (M) in kilograms by the moment of inertia of the load (J) in newton-meter² and the spring constant (K) by the torsional stiffness of the shaft assembly, also designated by (K) in newton-meters/radian, and the damping (f) in the spring–mass system by the rotational damping (B) in newtons-meters/radian per second, we end up with the second-order differential equation for the rotary system:

$$\theta_{(s)} = \frac{K/J}{s(s^2 + (B/J)s + K/J)} = \frac{\omega_n^2}{s(s^2 + 2\zeta\omega_n \cdot s + \omega_n^2)} \qquad (10\text{-}33)$$

from which

$$s^2 + \frac{B}{J} \cdot s + \frac{K}{J} = s^2 + 2\zeta\omega_n \cdot s + \omega_n^2 \qquad (10\text{-}34)$$

$$\omega_n^2 = \frac{K}{J} \quad \text{or} \quad \omega_n = \sqrt{\frac{K}{J}} \qquad (10\text{-}35)$$

and

$$2\zeta\omega_n = \frac{B}{J} \quad \text{or} \quad \zeta = \frac{B}{2\omega_n \cdot J} \qquad (10\text{-}36)$$

where

$$\frac{J}{B} = \tau, \text{ the time constant*} \qquad (10\text{-}37)$$

Also,

$$\zeta = \frac{B}{(2J)\sqrt{K/J}} \quad \text{and} \quad \frac{B}{2\sqrt{KJ}} \qquad (10\text{-}38)$$

and for critical damping,

$$\zeta = 1 = \frac{B}{2\sqrt{KJ}} \quad \text{and} \quad B = 2\sqrt{KJ} \qquad (10\text{-}39)$$

10-5 THE BASIC SERVO SYSTEM

Using the same technique, the closed-loop servo system can be quickly analyzed. Given the transfer function for a closed-loop system as

$$\text{TF} = \frac{K_o}{s(s\tau_m + 1) + K_o} \quad \text{or} \quad \frac{K_o}{s^2\tau_m + s + K_o} \qquad (10\text{-}40)$$

we divide the numerator and the denominator by τ_m:

$$\text{TF} = \frac{K_o/\tau_m}{s^2 + (1/\tau_m)s + K_o/\tau_m} \qquad (10\text{-}41)$$

For $\omega_n^2 = K_o/\tau_m$ and $1/\tau_m = 2\zeta\omega_n$ [see equations (8-17) and (8-18)], equation (10-41) can be rewritten as

$$\text{TF} = \frac{\omega_n^2}{s^2 + 2\cdot\zeta\omega_n\cdot s + \omega_n^2} \qquad (10\text{-}42)$$

Since equations (10-41) and (10-42) have the identical forms, then for a given system,

$$\frac{\omega_n^2}{s^2 + 2\zeta\omega_n\cdot s + \omega_n^2} = \frac{K_o/\tau_m}{s^2 + 1/\tau_m\cdot s + K_o/\tau_m} \qquad (10\text{-}43)$$

---

* Appendix A details the development of the motor time constant in terms of J and B.

and
$$s^2 + 2\zeta\omega_n \cdot s + \omega_n^2 = s^2 + \frac{1}{\tau_m} \cdot s + \frac{K_o}{\tau_m} \quad (10\text{-}44)$$

from which
$$\omega_n^2 = \frac{K_o}{\tau_m} \quad \text{and} \quad \omega_n = \sqrt{\frac{K_o}{\tau_m}} \quad (10\text{-}45)$$

$$2\zeta\omega_n = \frac{1}{\tau_m}$$

and
$$\zeta = \frac{1}{2\omega_n \tau_m} \quad (10\text{-}46)$$

Substituting equation (10-45) into equation (10-46) yields

$$\zeta = \frac{1}{2[(K_o/\tau_m)(\tau_m)^2]^{1/2}} = \frac{1}{2\sqrt{K_o \cdot \tau_m}} \quad (10\text{-}47)$$

For $\zeta = 1$, the case for critical damping:

$$K_o = \frac{1}{4\tau_m} \quad (10\text{-}48)$$

GLOSSARY

Critical Damping: The amount of damping that reduces the overshoot to zero when a resonant system is excited by a step input. It occurs when $\omega_d = \omega_n\sqrt{1 - \zeta^2} = 0$ or $\zeta = 1.0$.

Damping: The absorption of energy in an oscillatory system due to losses in the system.

Damping Coefficient: a—in an electrical circuit, the losses in the circuit that determine the rate of decay of a damped wave as a function of time $= a = R/2L$.

Damping Factor: $\zeta =$ the losses that determine the rate of decay of a damped wave as a function of frequency ω; taken as $\zeta = 1/(2\omega_n\tau_m)$.

Damped Frequency: The frequency that results when an oscillatory system is excited with a step input. It depends on the damping ζ and the natural frequency ω_n and is related as $\omega_d = \omega_n\sqrt{1-\zeta^2}$.

Dynamic System: One that is operating and performing a task.

Natural Frequency: In an electrical or mechanical resonant system, the steady-state frequency at which the phase angle of the output response is $-90°$ with respect to the input driving frequency. In an electrical system, it is $1/\sqrt{LC}$; in the spring–mass system, it is $\sqrt{K/M}$; and in a servo system, it occurs at $\sqrt{K_o/\tau_m}$. For $\zeta > 0$, ω_n is always higher than ω_d or ω_m.

Overshoot: The percentage ratio between the amount the output exceeds its final steady-state value divided by the final steady-state value with a step input command signal.

Response Time: The time it takes for the output to go from 10% to 90% of its final value with a step input as the command signal.

Settling Time: The time in seconds (taken as four time constants) for the output to fall within 2% of its final steady-state value with a step input as the command signal.

PROBLEMS

10.1   (a) What is the output frequency with a step input applied of an oscillatory system whose natural frequency is 5 Rad/s if $\zeta = 0.5$?

       (b) What happens to ω_d when ζ is increased to 0.707?

10.2   If $\omega_n = 25$ rad/s, and the damping coefficient $a = 12.0$:

       (a) What is the damped frequency of the system?

       (b) What happens to ω_d when a becomes 6.0?

       (c) What values of a will critically damp the system?

10.3   Prove $\omega_d = \omega_n\sqrt{1-\zeta^2}$ using $\omega_d = \sqrt{\omega_n^2 - a^2}$. (recall $Q = \omega_n L/R$ and $\zeta = 1/2Q$.)

10.4 (a) In a linear spring–mass system, determine the natural resonant frequency of the system when the spring compresses 0.01 m when subjected to a force of 1 N and the mass supported by the spring is 0.5 kg? (K = newtons/meter = $1/0.01 = 100$.)

(b) What value of damping (f) in newton-seconds/meter will provide a ζ of 0.3? [See equation (10-29)].

(c) What is the damped frequency when $\zeta = 0.3$? Give your answer in hertz.

10.5 If an automobile spring compresses one foot when it supports 1/2 ton of weight, what is ω_n of the structure? (*Hint:* A force of 1 lb. ≈ 4.3 N, a mass of 1 lb. ≈ 0.45 kg and 1 foot ≈ 0.3 m.)

10.6 (a) If the natural resonant frequency of a rotating spring–mass system is 10 rad/s, and $J = 50$ g-cm², what value of damping B (in dyn-cm/rad/s) is required for critical damping? [Use equations (10-36) and (10-39).]

(b) What is the time constant of the system for $\zeta = 0.3$?

10.7 A motor has a moment of inertia $J = 10$ g-cm² and viscous damping of $B = 50$ dyn-cm/rad/s:

(a) What is the time constant of the motor?

(b) If $K_o = 50$, determine ω_n when the motor is part of an automatic control system used for position control.

(c) What is the natural resonant frequency in hertz?

(d) Determine ω_d for this system when the loop is closed.

10.8 If the motor time constant of a closed-loop servo system is 0.6 s:

(a) What is the overall system gain that will result in good system response ($\zeta = 0.45$)?

(b) What happens to the damping if a higher K_o is used?

10.9 Given the closed-loop system with

$K_a = 50$

$K_{vm} = \dfrac{10}{s(0.06s + 1)}$ rad/s/V

K_{po} = to 2 V/rad

$N = 50 : 1$

- (a) What is the closed-loop transfer function?
- (b) What is the natural frequency ω_n of the system?
- (c) Convert your closed-loop equation to the steady-state form; what is the amplitude M_p at ω_p?
- (d) What is the overshoot with a step input?

10.10 Given the open-loop transfer function as, $\text{TF} = 50/[s(0.1s + 1)]$:
- (a) What is the natural resonant frequency ω_n?
- (b) What is ζ?
- (c) What is the % overshoot with a step input?
- (d) How many cycles are needed to reduce the output to under 2%?
- (e) Determine the amplitude of the 2nd peak of oscillation and check your answer by inspecting Figure 10-5.

11

Design Considerations of Dynamic Systems

11-1 INTRODUCTION

What follows is a summation of the fundamental concepts used in the design and development of automatic control systems. The overall approach is somewhat simplified and assumes conditions that are not always the case:

1. All reactions are linear.
2. The dynamic range of amplifiers is very large and saturation is not a problem.
3. Gear trains are perfect; they have no back lash and are matched to the load for the fastest response.
4. Dead space and velocity lag error are minimal.
5. The velocity limit* of the motor can be neglected.

* The maximum degrees/second that can be obtained from the output shaft of an automatic control system limited by the maximum motor speed obtainable.

Nevertheless, the concepts presented do provide for a good understanding of:

1. The language of automatic control systems.
2. Basic design principles.
3. The predictability of the behavior of a system under dynamic conditions.

11-2 THE MOTOR TRANSFER FUNCTION IN TERMS OF ITS SPEED/TORQUE CHARACTERISTICS

The key to setting up an automatic control system is the choice of the servo motor. A look at Figure 11-1 shows the typical characteristic curve of a small motor (the example of Figure 6-14). From Figure 11-1, the maximum power must occur at 50% of rated speed and torque, which means that the actual power developed at this point is 0.5×0.5, or 25% of the product of free speed and

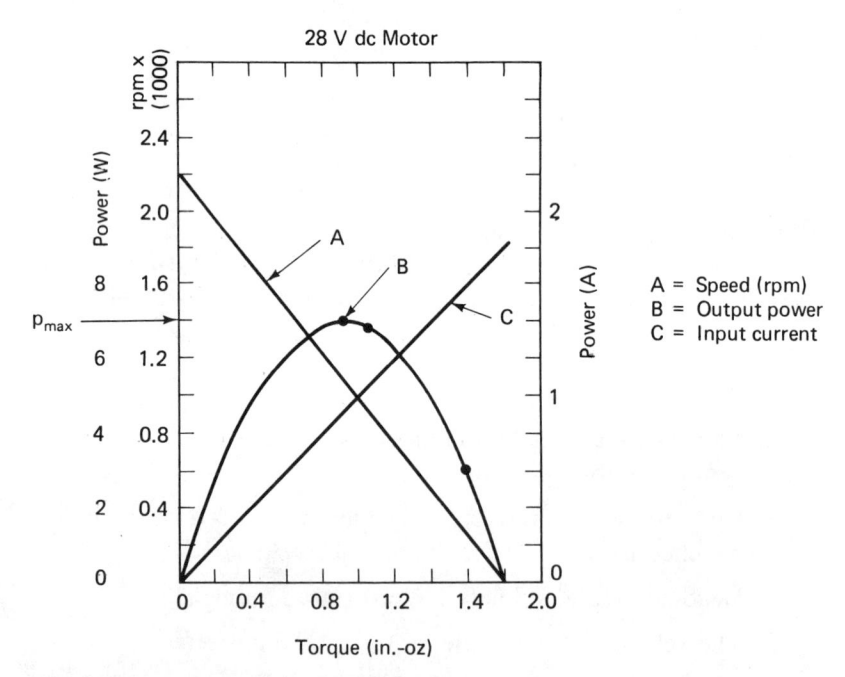

FIGURE 11-1 Characteristic curves of a servo motor (dc), size 8. (Courtesy of The Singer Company, Kearfott Division.)

Sec. 11-2 The Motor Transfer Function in Terms of its Speed/Torque   205

stall torque. This is because work is force × distance and at free speed there is zero force and at stall there is zero speed, making the output power zero at no load and at stall. For design purposes, the product of speed and torque of the load must be less than the maximum power output of the motor. Knowing the speed at maximum power output ($\frac{1}{2}$ free speed), the choice of the gear ratio* becomes a matter of the velocity (rad/s) desired of the final output of the system and the torque requirements of the load.

For example, from the data of Figure 11-1, determine the maximum power that the motor can deliver.
Given

$$P = T \cdot \theta\tau \qquad (11\text{-}1)$$

Where $P =$ horsepower

$T =$ torque in foot-pounds

$\theta\tau =$ angular velocity in rad/min

(recall 1 hp $=$ 33,000 foot-pounds/min and 1 hp $=$ 746 watts then:

$$P_{(\text{watts})} = \text{joules/s} = \frac{\frac{\text{in} - \text{oz}}{(12)(16)}(2\pi \text{ rpm})(746 \text{ W/hp})}{33{,}000 \text{ foot } \#/\text{min}}$$

$$P = \frac{(\text{in} - \text{oz})(\text{rpm})}{1.352 \times 10^3} \text{ joules/s} \qquad (11\text{-}2)$$

Since maximum power occurs at 1/2 maximum speed (2200/2 rpm), and the torque at this speed is 0.9 in $-$ oz:

$$P = \frac{(0.9)(2200/2)}{1.352 \times 10^3} = 0.732 \text{ joules/s}$$

11-2-1 Motor and Load Considerations

In Chapter 1, we discussed motor response as a function of time and know that the instantaneous speed of the motor at any time (t) is

$$\omega_o = S(1 - e^{-t/\tau}) \qquad (11\text{-}3)$$

---

* A good coverage of the subject appears in Chapter 5 of *The Servo Engineer's Handbook* (Weston Components Division, Weston Transcoil, Worcester, Pa.

where ω_o = instantaneous speed
S = no-load speed
$\tau = J/B$

and the TF in terms of s^* is

$$\theta_o/E_i = \frac{K_{vm}}{s(sJ/B + 1)} \qquad (11\text{-}4)$$

where J = moment of inertia
B = viscous damping
K_{vm} = rad/s/V
E_i = step input voltage
s = complex quantity = $0 + j\omega$ or simply $j\omega$

In pictorial form, a motor and gear train driving an external load can be depicted as shown in Figure 11-2. Applying input

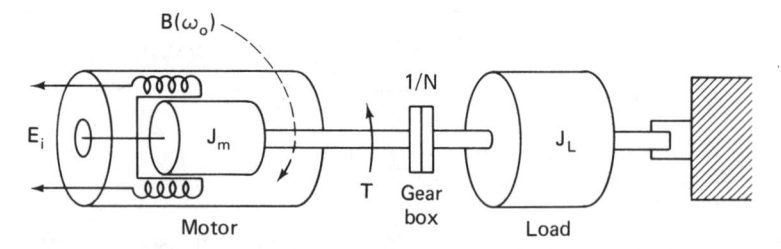

FIGURE 11-2 Pictorial drawing showing motor/load relationship.

voltage to the motor (E_i) creates a torque (T) which rotates the rotor, producing a polar moment of inertia (J_{motor}). As the motor speed builds up, the torque decreases until at free speed the torque produced is reduced to zero. The countertorque, which increased as ω_o increased, is referred to as B or viscous damping and is proportional to angular velocity and therefore to the counter EMF. We recognize N as a gear ratio and that the load has its own polar moment of inertia and damping, which is coupled to the motor through the gears. Just as in the case of the transformer with a turns ratio N, the total moment of inertia including the load

* See Appendix A

Sec. 11-2 The Motor Transfer Function in Terms of its Speed/Torque 207

reflected through the gear ratio N is modified by N^2 to:

$$J_T = J_m + J_L/N^2 \qquad (11\text{-}5)$$

For purposes of simplification, call the total moment of inertia J and the instantaneous viscous damping simply B. Then, according to equation (10-37), $\tau_m = J/B$ seconds. Both the moment of inertia (J) and the viscous damping (B) can be measured. It is a much simpler matter to measure the motor time constant (τ_m) directly. The example that follows is the actual plot of the output speed of a small servo motor under load which will be used in a model of a position control system to verify the conclusions reached in our classical analysis of servo-system behavior.

The circuit of Figure 11-3 was breadboarded and the output with a step input applied (by closing S_1) produced the response of Figure 11-4. The time constant τ_m is found from Figure 11-4 by

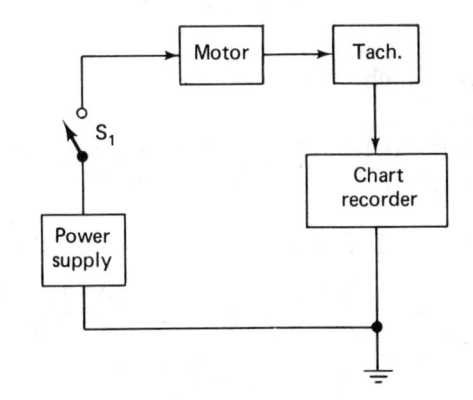

FIGURE 11-3 Breadboard circuit used to find τ_m.

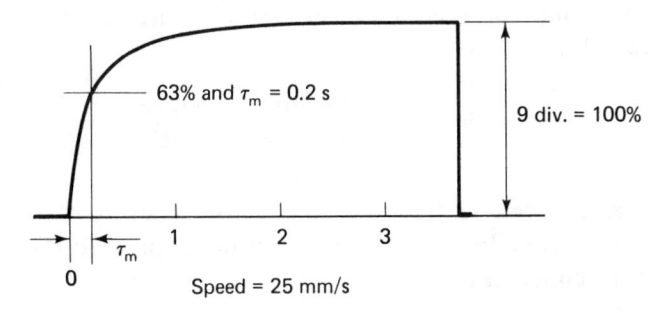

FIGURE 11-4 Chart recording of tachometer output of the circuit of Figure 11-3.

locating the time it takes to reach 63% of the final speed, which is shown to be approximately 0.2 s. Combining the data from Figures 11-1 and 11-4, a summary of the motor/load characteristics indicates:

1. Velocity constant:

$$K_{vm} = \frac{\text{no-load speed}}{\text{applied voltage}} = \frac{\text{rad/s}}{\text{V}}$$

$$= \frac{2200 \text{ rpm}}{28 \text{ V}} \cdot \frac{2\pi}{60} = 8.23 \text{ rad/s/V} \quad (11\text{-}6)$$

2. Torque constant:

$$K_{tm} = \frac{\text{max torque}}{\text{rated V}} = \frac{\text{dyn-cm}}{\text{V}} \quad (11\text{-}7)$$

To convert inch-oz to dyn-cm:

multiply inches by 2.54 to obtain centimeters
multiply ounces by 2.78 × 10⁴ to obtain dynes

$$2.54 \times 2.78 \times 10^4 = 7.06 \times 10^4 \text{ dyn-cm/in-oz}$$

$$K_{tm} = \frac{1.8 \times 7.06 \times 10^4}{28 \text{ V}} = \frac{12.71 \times 10^4}{28}$$

$$= 4.54 \times 10^3 \frac{\text{dyn-cm}}{\text{V}}$$

3. We could proceed to determine the viscous damping B which is:

$$B = \frac{\text{stall torque}}{\text{no-load speed}} = \frac{\text{dyn-cm}}{\text{rad/s}} \quad (11\text{-}8)$$

and proceed to find the time constant as $\tau_m = J/B$. The result from Figure 11-2 is sufficient to provide the motor time constant as:

$$\tau_m = 0.2 \text{ s}$$

11-3 BASIC SERVO SYSTEM WITH 100% FEEDBACK

A position-control servo system is to be designed using the motor characteristics stated above. It will be used with a reduction gear box with $N = 30:1$ and a set of position potentiometers that can be rotated through about 6 rad (344°) and are excited with ± 15 V to provide a sensitivity of

$$K_{pi} = K_{po} = \pm 15 \text{ V}/6 \text{ rad} = 5 \text{ V/rad}$$

The servo motor is driven by an amplifier with an effective gain of 0.9. The motor startup voltage is measured to be ± 0.75 V. The area of no response is the dead space which is equal to

$$\text{dead space} = \frac{1.5 \text{ V}}{5 \text{ V/rad}} = 0.3 \text{ rad or } 17.2°.$$

OBJECTIVE 1

Design a servo system using the components listed above with an adjustable preamplifier to control the overall gain of the complete system. The system in block form appears as shown in Figure 11-5.

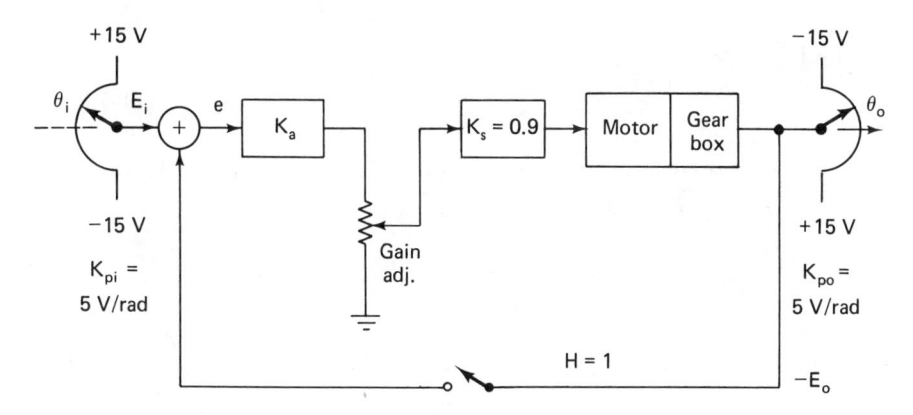

FIGURE 11-5 Block diagram of a position control system.

PROCEDURE

To obtain a fast response, a $\zeta = 0.4$ will be used. Determine the gain of the amplifier that will provide a damping factor of $\zeta = 0.4$ with $\tau_m = 0.2$ s and a K_{vm} of 8.23 rad/s/V.

209

The closed-loop transfer function based on the motor specified is

$$\text{TF}_{\text{CL}} = \frac{K_o}{s(0.2s + 1) + K_o} = \frac{K_o}{0.2s^2 + s + K_o} \cdot \frac{1/0.2}{1/0.2}$$

$$= \frac{5K_o}{s^2 + 5s + 5K_o} \tag{11-9}$$

This is in the form we recognize as a second order system resulting in:

$$\frac{5K_o}{s^2 + 5s + 5K_o} = \frac{\omega_n^2}{s^2 + 2\zeta\omega_n \cdot s + \omega_n^2} \tag{11-10}$$

From equation (11-10):

$$2\zeta\omega_n = 5$$

For $\zeta = 0.4$,

$$\omega_n = \frac{5}{2\zeta} = \frac{5}{0.8} = 6.25 \text{ rad/s} \tag{11-11}$$

Also,

$$\omega_n = \sqrt{\frac{K_o}{\tau_m}} \quad \text{and} \quad \omega_n^2 = \frac{K_o}{\tau_m}$$

and

$$K_o = \omega_n^2 \cdot \tau_m = (6.25)^2(0.2) = 7.81 \tag{11-12}$$

For the closed-loop system of Figure 11.5, the gain is defined as $K_o = E_o/e = E_o/E_i$ and for $E_i = 1$,

$$K_o = K_a \cdot K_s \cdot K_{vm}\left(\frac{1}{N}\right) \cdot K_{po}$$

Knowing that $K_o = 7.81$, K_a can be found.

$$7.81 = \frac{K_a(0.9)(8.23)(5)}{30}$$

and

$$K_a = \frac{7.81 \times 30}{(0.9)(8.23)(5)} = 6.33 \tag{11-13}$$

Sec. 11-3   Basic Servo System With 100 % Feedback   **211**

With the amplification ahead of the servo motor $= K_a \cdot K_s$, the dead space for this system is reduced by this added gain to

$$\text{dead space} = \frac{17.2°}{(6.33)(0.9)} = 3° \text{ total} \tag{11-14}$$

Next, substituting values into Figure 11-5, the block diagram becomes that of Figure 11-6. The complete open-loop transfer

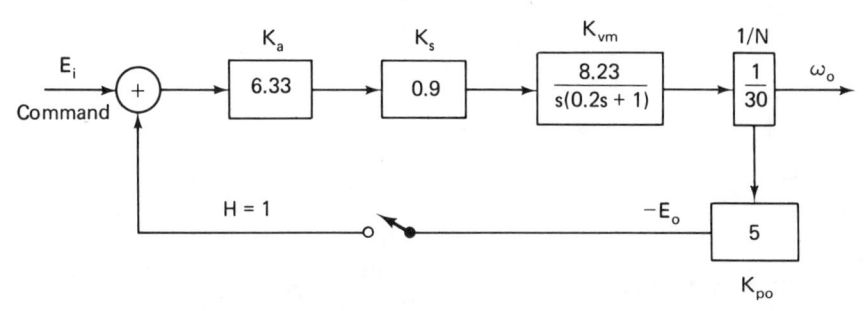

FIGURE 11-6   System of Equation 11-3 where $TF_{OL} = \frac{7.81}{[s(0.2s+1)]}$.

function for the system of Figure 11-5 shown in block form in Figure 11-6 is

$$TF_{OL} = \frac{K_o}{s(0.2s+1)} = \frac{7.81}{s(0.2s+1)} \tag{11-15}$$

The open-loop Bode plot of equation (11-15) is shown in Figure 11-7, from which

$$\omega_x = 5.5 \text{ rad/s}$$
$$\theta_x = -138°, \quad \text{providing a phase margin of } 42°.$$

To verify our data, substituting ω_x into equation (11-15) should give $A/\underline{\theta} = 0 \text{ dB}/\underline{-138°}$

$$A/\underline{\theta} = \frac{K_o}{s(0.2s+1)} = \frac{K_o}{j\omega(0.2j\omega+1)} = \frac{K_o}{-0.2\omega^2 + j\omega} \tag{11-16}$$

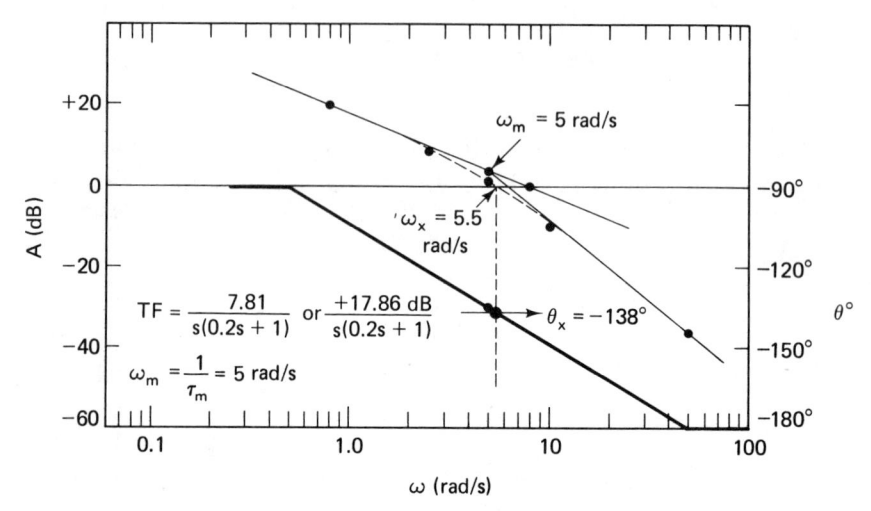

212 *Design Considerations of Dynamic Systems* Ch. 11

FIGURE 11-7 Open-loop plot of TF = $\dfrac{7.81}{s(0.2s+1)}$ with −3dB correction at ω_m.

Substituting $\omega_x = 5.5$ and $K_o = 7.81$,

$$A\underline{/\theta} = \frac{7.81}{-0.2(5.5)^2 + j5.5}$$
$$= \frac{7.81}{-6 + j5.5} = -0.36 \text{ dB}\underline{/-137.5°} \quad (11\text{-}17)$$

The phase margin of 42.5° is adequate for stable closed-loop performance.

Continuing with the closed-loop analysis, with $\omega_n = 6.25$ rad/s:

$$\omega_d = \omega_n\sqrt{1-\zeta^2}$$
$$= 6.25 \cdot \sqrt{1-(0.4)^2}$$
$$\omega_d = 5.73 \text{ rad/s}$$

The overshoot with a step input can be found, using equation (10-11), as

$$A = e^{-\zeta\pi/\sqrt{1-\zeta^2}}$$

Sec. 11-3   Basic Servo System With 100 % Feedback   **213**

where $\zeta = 0.4$

$$\pi/\sqrt{1-\zeta^2} = \frac{3.142}{\sqrt{1-(0.4)^2}} = 3.43$$

and

$$A = e^{-0.4(3.43)} = 0.25 \quad \text{or} \quad 25\%$$

Under steady-state conditions, with the loop closed,

$$\begin{aligned} \omega_p &= \omega_n\sqrt{1-2\zeta^2} \\ &= 6.25\sqrt{1-2(0.4)^2} \\ &= 6.25\sqrt{0.68} \\ &= 5.15 \text{ rad/s} \end{aligned} \tag{11-18}$$

The amplitude of the output at this frequency is found using

$$\begin{aligned} \underline{A/\theta} &= \frac{K_o}{s(s\tau_m+1)+K_o} = \frac{K_o}{-\omega^2\tau_m+j\omega+K_o} \\ &= \frac{7.81}{-(5.15)^2(0.2)+j5.15+7.81} \\ &= \frac{7.81}{2.51-j(5.5)} \end{aligned} \tag{11-19}$$

and

$$M_p = +2.22 \text{ dB}\underline{/-65.5°}$$

An analog model of the foregoing system (using a series of 741-C operational amplifiers) is shown in Figure 11-8:

The output response of the analog model to a step input (taken from a chart recorder) is shown in Figure 11-9. The measured overshoot is 26%, with a damped frequency of $\omega_d = 5.65$ rad/s. Incidently, the measured ω_p was 5.13 rad/s at a peak amplitude of +2.5 dB.

To complete our evaluation of the system of Figure 11-6, a laboratory model was set up and a chart recording was made of the output at K_{po}. It *exactly duplicated* Figure 11-9.

Here again, the overshoot was measured as 25% and ω_d

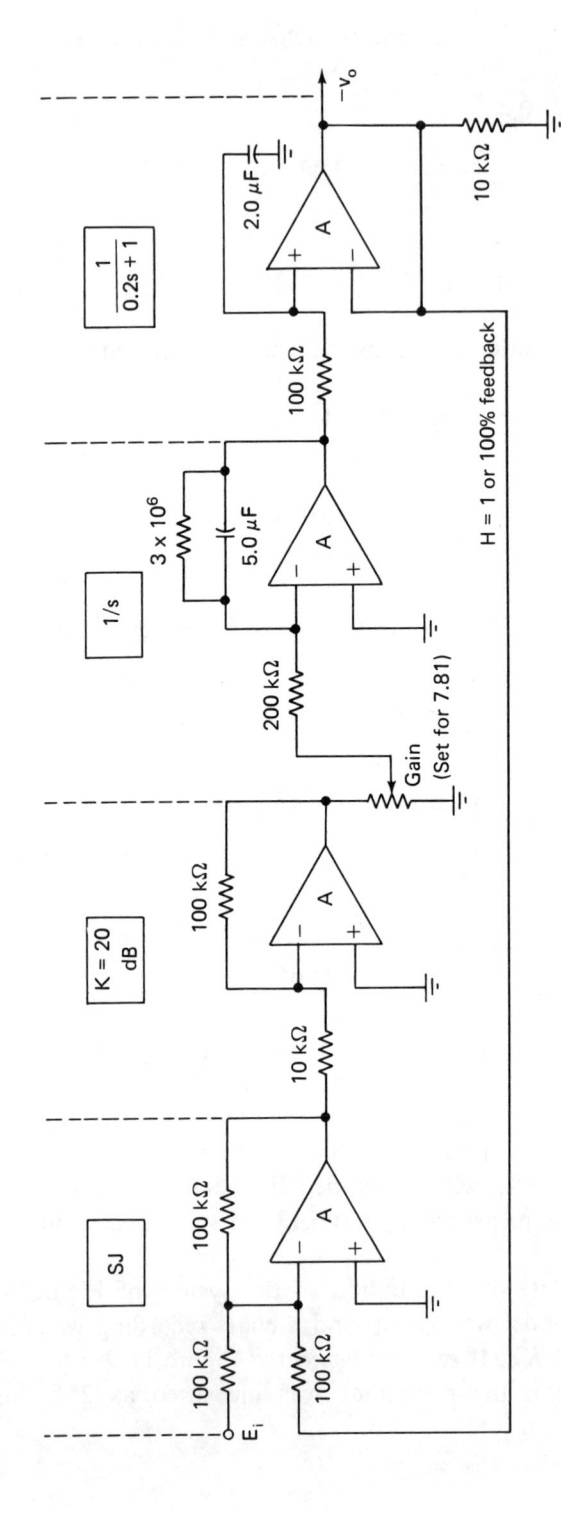

FIGURE 11-8 Analog model for $TF_{CL} = \dfrac{7.81}{[s(0.2s+1)+7.81]}$.

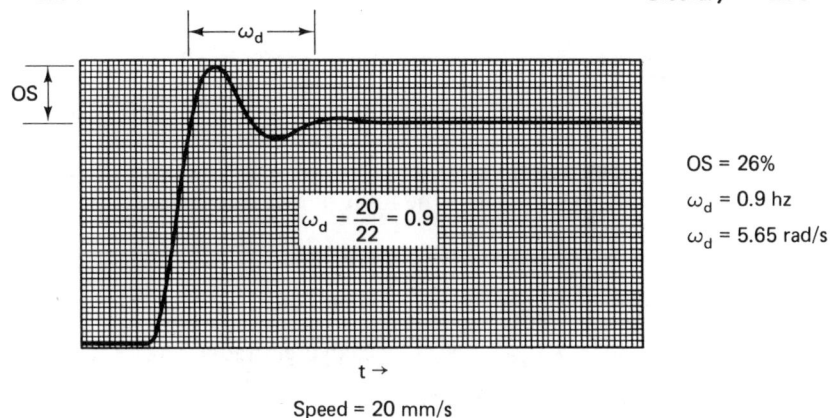

FIGURE 11-9 $TF_{CL} = \dfrac{7.81}{[s(0.2s + 1) + 7.81]}$ (analog model response to step input).

OS = 26%
ω_d = 0.9 hz
ω_d = 5.65 rad/s

$\omega_d = \dfrac{20}{22} = 0.9$

Speed = 20 mm/s

determined to be approximately 5.65 rad/s. The results of the mathematical analysis, the analog model, and the breadboard of the system described by equation (11-15) verify that the performance of a basic automatic control system can be predicted quite accurately by the techniques we have described in this text.

GLOSSARY

Analog Computer: A series of operational amplifiers with various forms of feedback to electronically simulate an electrical or mechanical system.

Moment of Inertia (J): A rotational inertia due to a moving mass as a function of its distance from the axis; mathematically, torque/acceleration expressed as g-cm^2.

Torque (T): The tendency of a force to produce angular acceleration, expressed as dyn-cm.

Viscous Damping (B): A countertorque proportional to angular velocity (ω), expressed in $\dfrac{\text{dyn-cm}}{\text{rad/s}}$.

PROBLEMS

11.1 The data given are already converted to proper units and can be used directly. A motor has the following characteristics:

$$J_m = 20 \text{ g-cm}^2$$
$$\text{no-load speed} = 5000 \text{ rpm}$$
$$B = 80 \text{ dyn-cm/rad/s}$$
$$\text{control voltage} = 25 \text{ V}$$

(a) What is the transfer function of the motor in a position control application?

(b) If the motor is driven by an amplifier with a gain of 20, what is the dead space if the input potentiometer has a sensitivity of 4 V/rad? (Assume that is start-up voltage for the motor of ± 1.5 V.)

11.2 The motor data given are:

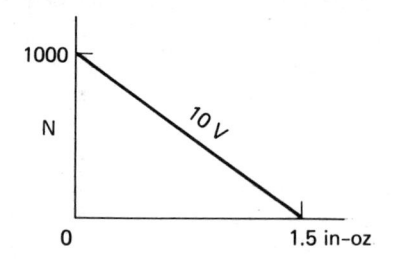

(a) Determine K_{vm}.

(b) Determine K_{tm} in dyn-cm/V.

(c) Determine $B = \dfrac{\text{stall torque}}{\text{(no-load speed)}} = \dfrac{\text{dyn-cm}}{\text{rad/s}}$

(d) If $J_m = 200$ gm-cm^2, write the transfer function of the motor.

(e) At what speed is maximum output power developed?

(f) What is the maximum power developed (in horsepower)?

11.3 The motor of Problem 11-2 is set into a system with: $K_{po} = 4$ V/rad, $N = 30:1$

      (a) If we desire $\zeta = 0.5$, determine the gain K_a required.

      (b) Determine the open-loop transfer function.

11.4 Prepare a corrected Bode plot of the transfer function developed for Problem 11-3, then on a Nichols chart:

      (a) Determine ω_n.

      (b) Determine ω_p and M_p.

11.5 (a) Prove your answers to Problem 11.4 using the transfer function of Problem 11-3.

11.6 What is the overshoot that can be expected with a step input as the command signal?

11.7 Lay out an analog computer model of the system using operational amplifiers with appropriate feedback.

Appendix A

The Motor Transfer Function in Terms of Moment of Inertia and Viscous Damping

In Chapter 1, we discussed motor response as a function of time and know that the instantaneous speed of the motor at any time (t) is

$$\omega_0 = \frac{d\theta}{dt} = S(1 - e^{-t/\tau}) \qquad \text{(A-1)}$$

where S = no-load speed
$\tau = J/B$ = time constant of the motor and load combined
$d\theta/dt$ = instantaneous speed ω_0
J = moment of inertia
B = viscous damping
θ_o = angular displacement

If θ_o is angular displacement in degrees, then $d\theta/dt$, the rate of change of displacement over change of time, must equal angular velocity (ω_0) as indicated by equation A-1.

Continuing with Newton's law, the first derivative of displacement is velocity and the change of velocity with time, which is the second derivative of displacement (θ), is acceleration (α).

In the same way that force $(F) = \text{mass} \cdot \alpha$ in a linear system, the torque or momentum (T) in a rotary system is

$$T = J_m \cdot \alpha \qquad \text{(A-2)}$$

where J_m = moment of inertia of the motor and load

$\alpha = d^2\theta/dt$

Substituting $d^2\theta/dt$ for α in equation (A-2) yields

$$T = J_m \cdot \frac{d^2\theta}{dt} \qquad \text{(A-3)}$$

An understanding of these basic laws of physics will help us to understand the development of our key equation for the open- and closed-loop transfer function of a servo system in terms of $J/B = \tau_m$.

A pictorial presentation of a motor driving a load through a gear-reduction box is depicted in Figure A-1. From Figure A-1,

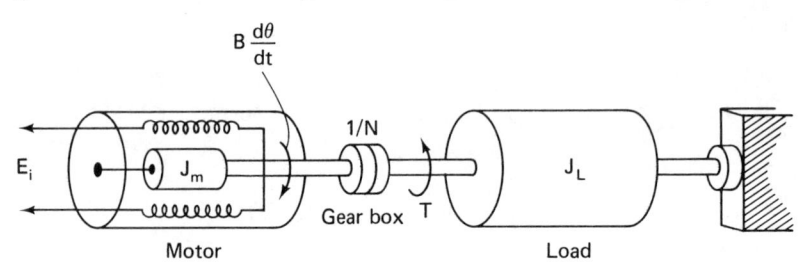

FIGURE A-1 Motor/load, pictorial presentation.

as the motor speed increases, the available output torque decreases until at full rated speed (when the acceleration is zero), the torque is zero.

$$T(\text{available}) = T(\text{stall}) - B \cdot \frac{d\theta}{dt} = 0 \qquad \text{(A-4)}$$

where T = torque

B = stall torque/rated speed

$d\theta/dt$ = instantaneous speed

Several of the other motor/load constants are recognized to be:

1. The velocity constant:

$$K_{vm} = \frac{\text{no load speed}}{\text{applied voltage}} = \frac{\text{rad/s}}{\text{V}} \quad (A\text{-}5)$$

2. The torque constant:

$$K_{tm} = \frac{\text{stall torque}}{\text{rated voltage}} = \frac{\text{dyn-cm}}{\text{V}} \quad (A\text{-}6)$$

3. There remains

$$\frac{\text{stall torque}}{\text{rated speed}} = \frac{\text{dyn-cm/V}}{\text{rad/s/V}} = \frac{\text{dyn-cm}}{\text{rad/s}} \quad (A\text{-}7)$$

Note that equation (A-7) has the exact units used to define B, viscous damping:

$$B = \frac{\text{stall torque}}{\text{rated speed}} = \frac{\text{dyn-cm}}{\text{rad/s}} \quad (A\text{-}8)$$

Rewriting equation (A-4) and replacing stall torque by $K_{tm} \cdot E_i$ from equation (A-6):

$$T = K_{tm} \cdot E_i - B\frac{d\theta}{dt} \quad (A\text{-}9)$$

Substituting our derivation for T from equation (A-3),

$$J\frac{d^2\theta}{dt^2} = K_{tm} \cdot E_i - B\frac{d\theta}{dt} \quad (A\text{-}10)$$

In the s domain, $d\theta/dt = s \cdot \theta$ (see Table F-2), and in terms of s, equation (A-10) becomes

$$Js^2 \cdot \theta = K_{tm} \cdot E_i - B \cdot s\theta$$

and

$$K_{tm} \cdot E_i = Js^2\theta + B \cdot s\theta = \theta(s^2 J + sB)$$

from which

$$\frac{\theta_o}{E_i} = \frac{K_{tm}}{s^2 J + sB} \quad \text{or} \quad \frac{K_{tm}/B}{s^2 J/B + sB/B} \quad (A\text{-}11)$$

Factoring s in the denominator and replacing K_{tm}/B by K_{vm} and $J/B = T_m$, equation (A-11) can be rewritten as

$$\text{TF} = \frac{\theta_o}{E_i} = \frac{K_{vm}}{s(sJ/B + 1)} \quad \text{or} \quad \frac{K_{vm}}{s(s\tau_m + 1)} \qquad \text{(A-12)}$$

Of course, equation (A-12) can be varied by adding various gains as desired, and as we now recognize, produces the key equation for a *closed-loop system*, where

$$\frac{\theta_o}{\theta_i} = \frac{K_o}{s(s\tau_m + 1) + K_o} = \frac{K_o}{s^2\tau_m + s + K_o}$$

and

$$\frac{\theta_o}{\theta_i} = \frac{K_o/\tau_m}{s^2 + 1/\tau_m \cdot s + K_o} \qquad \text{(A-13)}$$

With $K_o/\tau_m = \omega_n^2$ and $1/\tau_m = 2\zeta\omega_n$, equation (A-13) becomes

$$\text{TF}_{\text{CL}} = \frac{\theta_o}{\theta_i} = \frac{\omega_n^2}{s^2 + 2\zeta\omega_n \cdot s + \omega_n^2} \qquad \text{(A-14)}$$

The correlation between mechanical and electrical behavior in feedback systems is now complete.

Appendix B

Rate Feedback

In an earlier introduction to phase compensation (Chapter 6), we briefly discussed generator (or tachometer) feedback. Actually, the approach presented was greatly oversimplified. The analysis of % of rate feedback on system response is more complex and involves a two-feedback-loop analysis.

Figure B-1 is a typical Type 1 position servo system with the transfer function containing the $(1/s)$ or single integration. With

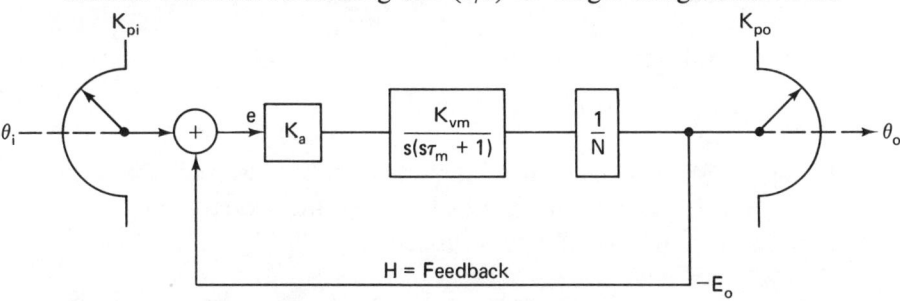

FIGURE B-1 Type 1 system = $K_o/S(S\tau_m + 1)$.

224 *Rate Feedback* App. B

the loop closed and $H = 1$ and $K_{po} = K_{pi}$:

$$\text{TF} = \frac{K_o}{s(s\tau_m + 1) + K_o(1)} \tag{B-1}$$

Adding variable rate feedback (also called velocity feedback), the system of Figure B-1 becomes a variation of the Type 1 system or that of Figure B-2. Normally, the tachometer is connected

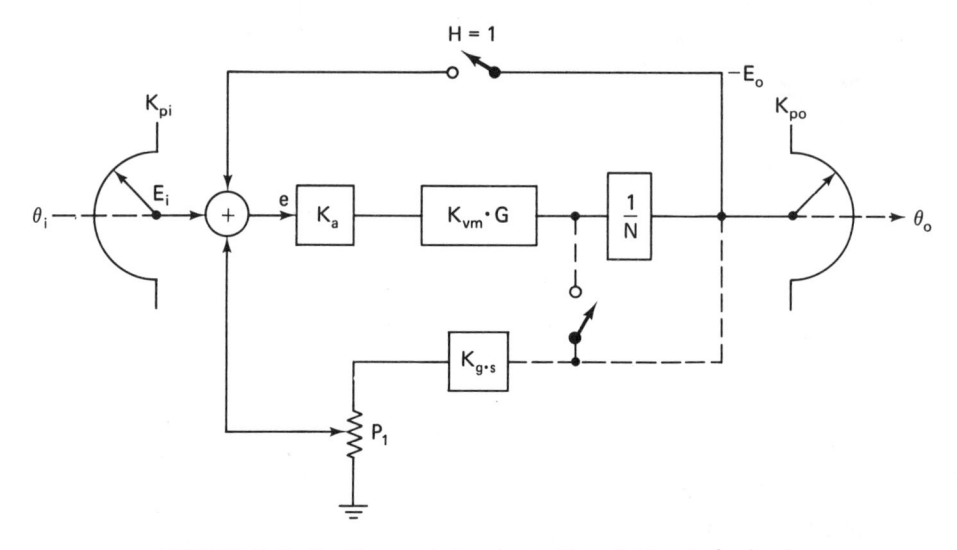

FIGURE B-2 Position control system with variable rate feedback.

directly to the output shaft of the motor (shown by dashed lines). Since P_1 sets the amount of rate feedback desired, we will retain the $1/N$ factor in the equation to simplify the mathematical derivation that follows. Note there are two feedback loops:

1. Overall feedback with $H = 1$
2. Rate feedback (H_2) governed by P_1

In Figure B-2, with the outer loop open, the second feedback loop which contains the rate generator and the potentiometer P_1 provides the amount of rate feedback as:

$$H_2 = K_g \cdot s \cdot P_1 \tag{B-2}$$

App. B   *Rate Feedback*

where $K_g \cdot s$ = TF of rate generator
               P_1 = % of $K_g \cdot s$ feedback

Replacing $H = 1$ with $H_2 = K_g \cdot s \cdot P_1$ in equation (B-1)

$$\text{TF}_{\text{OL}} = \frac{K_o}{s(s\tau_m + 1) + K_o \cdot K_g \cdot s \cdot P_1} \quad \text{(B-3)}$$

For ease in handling, combine $K_o \cdot K_g$ as K_R making

$$\text{TF} = \frac{K_o}{s(s\tau_m + 1) + s(K_R \cdot P_1)} \quad \text{or} \quad \frac{K_o}{s(s\tau_m + 1 + K_R \cdot P_1)} \quad \text{(B-4)}$$

Setting $P_1 = 0$ (with the outer loop "open"), equation (B-4) reduces to:

$$\text{TF} = \frac{K_o}{s(s\tau_m + 1)} \quad \text{(B-5)}$$

which is the open-loop transfer function, as it should be.

Adjusting P_1 so that $K_R \cdot P_1 = 1$, equation (B-4) becomes

$$\text{TF} = \frac{K_o}{s(s\tau_m + 1 + 1)} = \frac{K_o}{s(s\tau_m + 2)} = \frac{K_o/2}{s(s\tau_m/2 + 1)} \quad \text{(B-6)}$$

Notice, in equation (B-6), that the gain is down by $\frac{1}{2}$ and the corner frequency has increased 2×. With $K_R \cdot P_1 = 2$, equation (B-4) becomes

$$\text{TF} = \frac{K_o/3}{s(s\tau_m/3 + 1)} \quad \text{(B-7)}$$

In equation (B-7) the overall gain is down by $\frac{1}{3}$, but the real significance is that the effective time constant has been reduced to $\frac{1}{3}$ of τ_m. The effective decrease of τ_m by a factor of three provides a new natural resonant frequency:

$$\omega_n = \sqrt{\frac{K_o/\tau_m}{3}} = \sqrt{\frac{3K_o}{\tau_m}} \quad \text{(B-8)}$$

The higher ω_n provides a higher ω_d. The response time described by the first overshoot of ω_d is thus speeded up by $1/\sqrt{3}$

or 57%. There is a limit to how far this technique can be used to speed up the response time. Since the effective $\tau_m = J/B$ is decreased, and J is a quantity determined by the physical properties of the system, then B, the viscous damping, is modified by adding rate feedback. For τ_m to decrease, B must increase. Referring to the troque equation (A-4) in Appendix A where:

$$T(\text{available}) = T(\text{stall}) - B \cdot d\theta \cdot dt \qquad (B-9)$$

As the effective B increases, the available torque decreases for a given output speed $d\theta/dt$ in which case the output begins to fall behind thereby increasing the velocity lag error in the system. The design specifications will determine the limit to which the viscous damping can be increased.

Example B-1

Using the transfer function from Chapter 9, equation (9-9), for a comparison, what happens to the system performance with rate feedback added? Assume that τ_m is modified to $\tau_m/3$. The original transfer function is

$$\text{TF}_{\text{OL}} = \frac{26.6}{s(0.05s + 1)} \qquad (B-10)$$

The Bode plot is shown in Figure B-3. An extension of the Bode plot is the modified transfer function with τ_m replaced by $\tau_m/3$. The transfer function becomes

$$\text{TF} = \frac{26.6}{s\left(\dfrac{0.05 + 1}{3}\right)} = \frac{26.6}{s(0.0167s + 1)}$$

where $\omega_{c_2} = 60$ rad/s

It becomes evident that the open-loop gain can be increased by $+9\ dB$ or 2.8 times, and still maintain the phase margin of 42° at the new gain crossover frequency. Thus, a new transfer function is generated:

$$\text{TF}_{\text{OL}} = \frac{26.6 \times 2.82}{s(0.0167s + 1)} \quad \text{and} \quad \text{TF}_{\text{CL}} = \frac{75}{s(0.0167s + 1) + 75}$$

App. B   Rate Feedback **227**

$$TF_1 = \frac{15.8}{s(0.05s + 1)}$$

$$TF_2 = \frac{15.8 \text{ or } 24 \text{ dB}}{s\left(\frac{0.05}{3}s + 1\right)}$$

$$TF_3 = \frac{+34 \text{ dB or } 50.1}{s(0.0167s + 1)}$$

$\omega_{c2} = 60$ rad/s

$\theta_x = 130°$ 3°/div.

$\phi_x = 50°$

FIGURE B-3 Bode plot of a transfer function with or without rate feedback.

$$TF = \frac{15.8}{s(0.05s + 1)}$$

Revised to:

$$TF = \frac{15.8}{\left(s\dfrac{0.05}{3}s + 1\right)}$$

$\omega_{c2} = 60$ rad/s

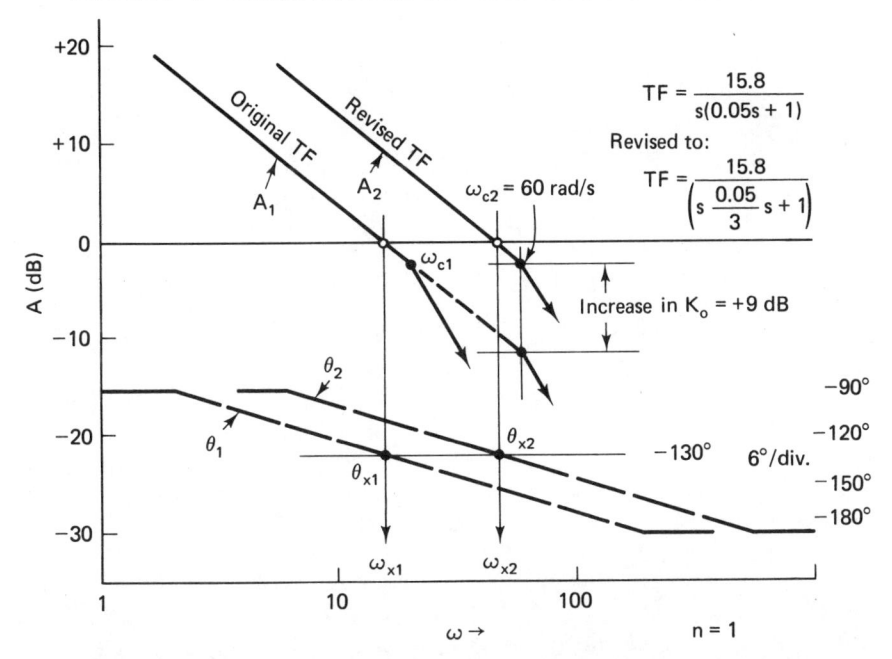

Increase in $K_o = +9$ dB

$-130°$ 6°/div.

n = 1

FIGURE B-4 Bode plot of Transfer function with and without rate feedback.

from which $\omega_n^2 = 75$ and $\omega_n^2 = 8.66$ rad/s. This compares with the original case, where $\omega_n = \sqrt{26.6}$ or 5.16 rad/s.

In summation, the addition of a small amount of rate feedback permitted a higher preamplifier gain, which decreased the dead space by a factor of almost 3 times. Also, the higher natural resonant frequency (ω_n) increased the damped resonant frequency (ω_d), which increases the response time by a factor of more than 1.5 x. In conclusion, rate feedback (also called velocity feedback) is very effective in improving system performance.

Appendix C

The Lead Network

Lag networks or lead networks can be used to improve the phase margin of an automatic control system. We will limit our network compensation to the lead type, since it is much simpler to analyze and the results that are obtained are quite effective.

The lead network, like the tachometer is a differentiator or a derivative device. Therefore, the effects obtained with a lead networks is quite similar to that obtained with rate feedback; namely, the system response is improved and with increased system gain, the "dead space" is decreased.

In Chapter 7, it was shown that a lead network is of the form

$$\text{TF} = \frac{1 + s\tau_1}{1 + s\tau_1/10} \tag{C-1}$$

This type of lead network will provide a $+45°$ phase angle (linearized) over a wide range of frequencies from $1/\tau_1$ to $10/\tau_1$. For the system of equation (7-1) given as

$$\text{TF} = \frac{100}{s(0.1s + 1)} \quad \text{where } 0.1 \text{ s} = \tau_m \tag{C-2}$$

the plot of amplitude (curve A) and phase (curve B) are given in Figure C-1. By making τ_1 of equation (C-1) equal to τ_m of equation (C-2), then over the range of frequencies $1/\tau_m$ to $10/\tau_m$, a $+45°$ phase response is obtained, as shown by curve C.

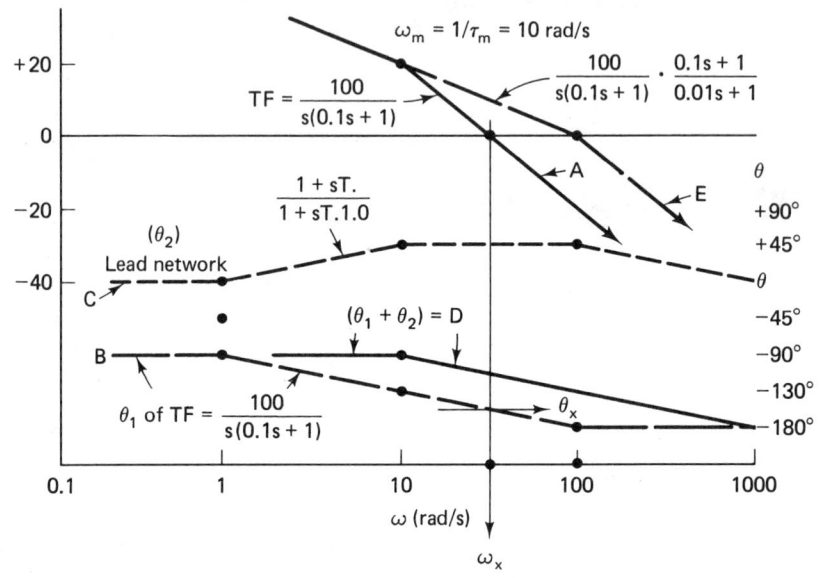

FIGURE C-1 Adding a lead network $\left(\dfrac{s\tau_m + 1}{s\tau_m/10 + 1}\right)$ to TF $= \dfrac{100}{s(0.1s + 1)}$ with $\tau_m = 0.1\text{s}$.

When curve C is added to the original phase plot of curve B, the result is curve D (the sum of $\theta_1 + \theta_2$). In addition to the affect on the total phase response, there is also a change in the amplitude characteristics. In fact, for the network of equation (C-1), the effective corner frequency is raised by a factor of 10 times and the gain crossover is moved to a much higher frequency, as seen by curve E. It should become apparent that as the corner frequencies are changed in the lead network, the gain crossover frequency is also shifted, and that the frequencies at which the $+45°$ is added to the overall response must move up or down. The combination of τ_1 and τ_2 that will result in an improved phase margin can vary from $\tau_1 = \tau_m$ to as high as $\tau_m/10$.

In Section 11-1, we listed several factors that can limit the accuracy of a design. In practice, a lead network can be a simple circuit whose frequency characteristics in terms of $(s\tau_1 + 1)/$

$(s\tau_2 + 1)$ are readily modified by varying the capacitor C or in some cases the resistor R_1 (Figure C-2). It is not a difficult procedure to optimize a particular servo system under dynamic conditions by adjusting the K_o and the value of C for maximum usable gain for the desired system-response characteristics.

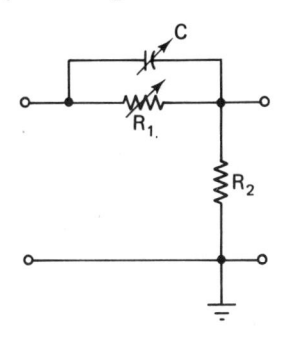

FIGURE C-2 Lead network where T_1 and T_2 can be varied.

For the systems discussed in the text, the values of τ_1 for the lead network varied from $\tau_m/2$ to as high as $\tau_m/6$ for the highest usable system gain with a good fast system response.

To summarize, the simple procedure of adding a lead network of the type shown in Figure C-2, with $\tau_1 = \tau_m$, results in:

1. A higher gain crossover frequency.
2. Permits at least a 20-dB or 10× increase in the system gain (K_o), which increases ω_n by a factor of $\sqrt{10}$, or 3.16 times ($\omega_n = \sqrt{K_o/\tau_m}$).
3. The dead space is reduced by the increase in the system gain by a factor of more than 3×.
4. The increased ω_n results in an increased value for ω_d. Since the response time is based on the first overshoot of ω_d, this is also increased by approximately a factor of $\sqrt{3}$ or 1.7×.

Finally, and of very real significance, is the fact that the phase margin is extended over a very large range of frequencies. This allows changes in system parameters, such as the gain or change in component values, not to drastically affect system performance once it is set.

Appendix D

Phase Margin

Establishing the phase margin is based on a set of compromises, the major one being the amount of overshoot that can be tolerated in the specific system. (Refer to the closed-loop response of Figure 10-5, which will clarify the effect of damping on overshoot.) For example, in many missile systems, where response time is important, the excessive wear on the moving parts that will result from too much overshoot is not a serious problem, since long life is not a criterion of design. In fact, a 25% overshoot corresponding to a damping factor of $\zeta = 0.4$ is not unusual. On the other hand, in many operations, such as a lathe cutting operation or systems designed for long life, any overshoot would shorten the life of the equipment considerably. Thus, a damping factor of $\zeta = 0.7$ or larger is required where the overshoot is zero.

The Nichols chart (Figure D-1) contains all the necessary open- and closed-loop data to readily define the phase angles that exist for the damping factor specified. From Table 8-3, for $\zeta = 0.4$, $M_p \approx +3$ dB, and for $\zeta = 0.7$, $M_p \approx 0$ dB. On the Nichols

234 Phase Margin **App. D**

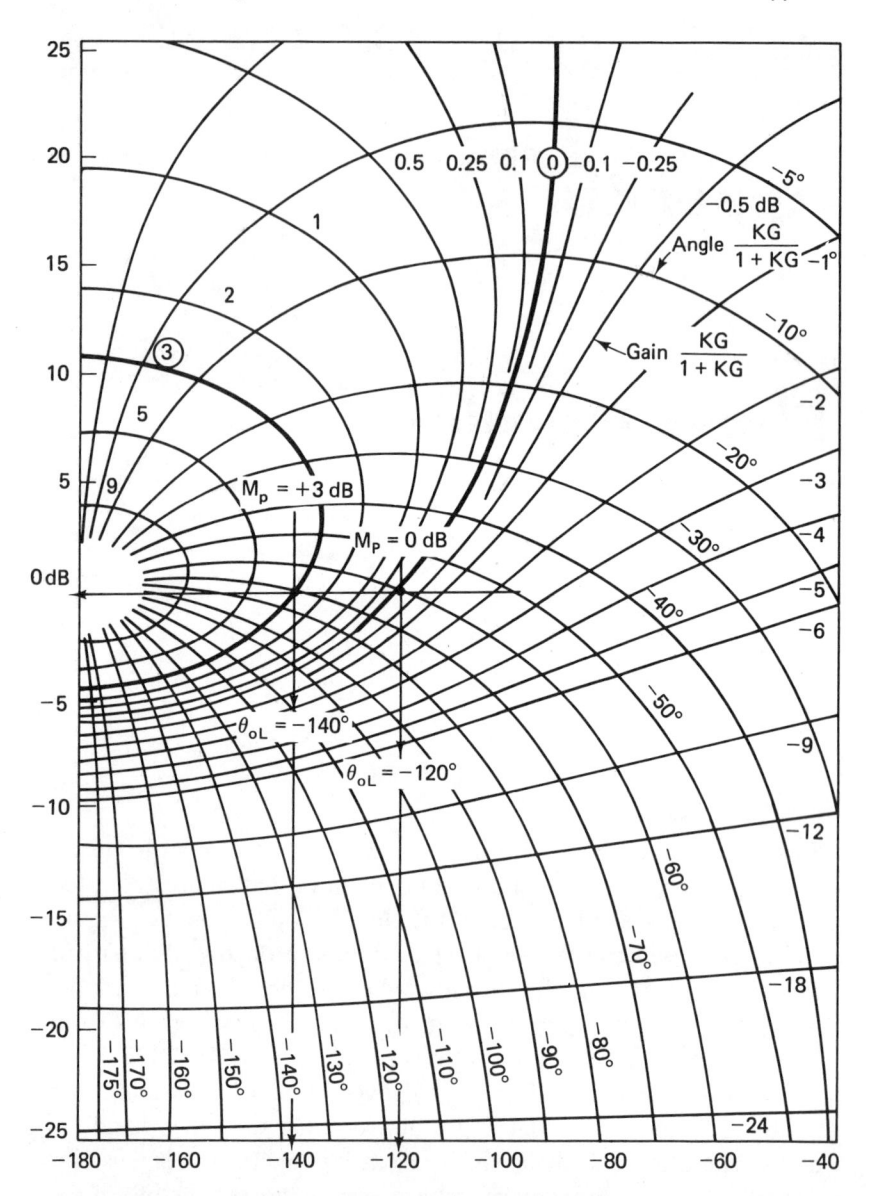

FIGURE D-1 Criteria for establishing the phase margin as 40° → 60° based on ζ of 0.4 to 0.7 resulting in a 0 to 25% overshoot.

chart, the point specifying the *closed-loop data* for $M_p = +3$ dB crosses the 0-dB line of the *open-loop* gain plot at ω_x (since, by definition, the open-loop gain at ω_x is 0 dB). The open-loop phase angle at ω_x is seen to be $\theta_{OL} = -140°$ and the phase margin is

40°. Similarly, with $M_p = 0$ dB, the heavy line of the *closed-loop data* provides the open-loop data at $A = 0$ dB (which by definition is at ω_x). The phase angle $\theta_{\text{OL}} = -120°$ and the phase margin (ϕ) is $(180° - 120°) = 60°$.

Based on the foregoing criteria, the phase margin must *not be less than 40°* or the overshoot and oscillations that result will be excessive. At the other limit, a phase margin *in excess of 60°* will result in an overdamped system that will be slow or sluggish in response to a command signal.

Appendix E

A Speed Control System

Let us replace the blocks of our position control system to create the speed control system of Figure E-1. In this system, an input voltage E_i is amplified to drive the servo motor. Directly coupled to the shaft of the servo motor is a tachometer whose output is V/rad/s. If the output of the tachometer is added to the input signal (out of phase), the error (e) goes toward zero, since $E_o - E_i = e$. With a tachometer sensitivity of 1 V output/100 rad/s, to obtain a steady speed of, say, 500 rad/s requires that $E_o = -5$ V and that E_i be set to $+5$ V. Any change in the output speed due to

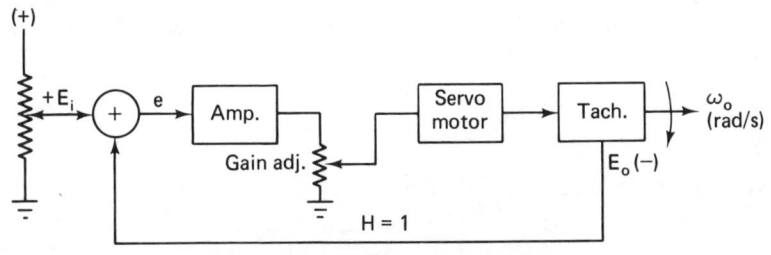

FIGURE E-1 Speed control system.

loading effects or line voltage changes results in a rise or fall of $-E_o$ above or below the reference $+5$ V, and the motor drive is increased or decreased as necessary.

This simple system is quite effective, and substituting transfer functions for the blocks shown in Figure E-1 results in the speed control system in Figure E-2. Of real significance in Figure E-2

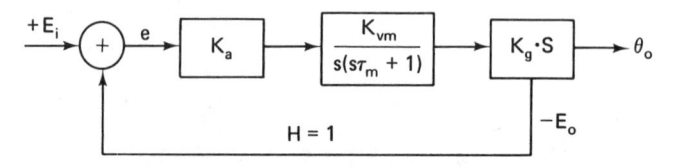

FIGURE E-2 Speed control system.

is that the output fed back to the input is the voltage output of the tachometer (TF = $K_g \cdot s$). In reality, a speed control system depends on the voltage regulation of the tachometer output. Thus, it acts like a closed-loop voltage amplifier with 100% feedback, and the objective is to maintain $E_o/E_i = 1$.

From Figure E-2, the output error is defined as

$$e = E_i - E_o \tag{E-1}$$

and the closed-loop gain is

$$G = \frac{E_o}{E_i} = \frac{A}{1 + A(B)} \tag{E-2}$$

Substituting $K \cdot G$ for A and $B = 1 = 100\%$ feedback:

$$TF = \frac{E_o}{E_i} = \frac{K \cdot G}{1 + K \cdot G} \tag{E-3}$$

with

$$K \cdot G = \frac{K_a \cdot K_{vm} \cdot K_g \cdot s}{s(s\tau_m + 1)} \tag{E-4}$$

The overall closed-loop transfer function for a speed control system becomes

$$TF = \frac{K_o}{(s\tau_m + 1) + K_o} \tag{E-5}$$

with an open-loop transfer function of

$$TF = \frac{K_o}{s\tau_m + 1} \tag{E-6}$$

Notice in equation (E-6) that the integration ($1/s$) of equation (E-4) is canceled. Since no ($1/s$) appears in the final transfer function, a speed control system becomes a Type 0 system, with the only time constant of significance being that of the motor.

At very low frequencies, a decade or so below the corner frequency $1/\tau_m$, the transfer function becomes

$$\text{TF} = \frac{K_o}{1 + K_o}$$

As with any closed-loop system with 100% feedback, as the forward gain is increased, E_o/E_i more closely approaches unity in which case $E_i = E_o$. Thus, the error, $E_i - E_o = e$ also approaches zero.

The transfer function of the tachometer in the s domain is given as

$$\text{TF} = \frac{E_o}{\theta_o} = K_g \cdot s$$

and

$$E_o = K_g \cdot s \cdot \theta \tag{E-7}$$

Replacing $s \cdot \theta$ with ω_0 (see pair 10 of Table 4-1), the transfer function in the time domain becomes

$$E_o = \omega_0 \cdot K_g$$

and

$$\omega_0 = \frac{E_o}{K_g} = \frac{\text{V}}{\text{V/rad/s}} = \text{rad/s} \tag{E-8}$$

Since E_o is closely regulated and K_g is a constant, ω_0 (rad/s) must be closely regulated.

With basically only one corner frequency affecting the system, there is very little likelihood of instability in a speed-control system of this type. The real problem is that of setting the amplifier gain. Too much gain will introduce jitter in the output speed from the noise generated within the system. Too little gain will result in a slow reaction time from insufficient drive to the servo motor when called upon to correct for a change in speed.

Appendix F

The Laplace Transformations

Mathematically, any function $f(t)$ is transformed into the s domain where $F(s)$ is the Laplace transform (sometimes referred to as the Laplace integral) by the use of equation (F-1), where

$$F(s) = \int_0^\infty f(t) \cdot e^{-st}\, dt \qquad \text{(F-1)}$$

The following examples will clearly demonstrate the transformation from $f(t)$ to $F(s)$ of several of the more basic time functions. With some assistance from the instructor, the student should be able to follow the development for his or her own enlightenment. Let us make it clear from the beginning that the derivation of many of the Laplace transforms can be difficult mathematical manipulations. Luckily for us, very complete tables of transforms* from $f(t)$ to $F(s)$ are available.

> * An excellent set of tables appear in Floyd E. Nixon, *The Handbook of Laplace Transformation*, 2nd ed. (Prentice-Hall, Inc., Englewood Cliffs, N.J., 1965).

F-1 EXAMPLES OF BASIC TRANSFORMS FROM $f(t)$ TO $F(s)$

EXAMPLE F-1

If $f(t)$ is A or any fixed quantity, a unit step quantity appears simply as

$$F(s) = \int_0^\infty A \cdot e^{-st}\, dt = A \int_0^\infty e^{-st}\, dt$$
$$= A(-1/s \cdot e^{-st})*$$

When $t = \infty$,

$$A\left(\frac{-1}{s \cdot e^\infty}\right) = \frac{-A}{s \cdot \infty} = 0$$

when $t = 0$,

$$-A\left(\frac{-1}{s \cdot e^0}\right) = \frac{+A}{s}$$

and for $A = 1$,

$$F(s) = \frac{1}{s}$$

Thus,

for the given $f(t) = A$, $F(s)$ is A/s \hfill (F-2)

EXAMPLE F-2

In a similar manner, take the popular time function $f(t) = e^{-at}$. Pictorially, $y = e^{-at}$ appears as shown in Figure F-1, where

$$f(t) = e^{-at} \text{ and } F(s) = \int_0^\infty e^{-at} \cdot e^{-st}\, dt = \int_0^\infty e^{-(s+a)t}\, dt \quad \text{(F-3)}$$

The solution is the same as that for Example F-1, where s is

---

* *Hudson Manual Table of Integrals*, p. 50, no. 303.

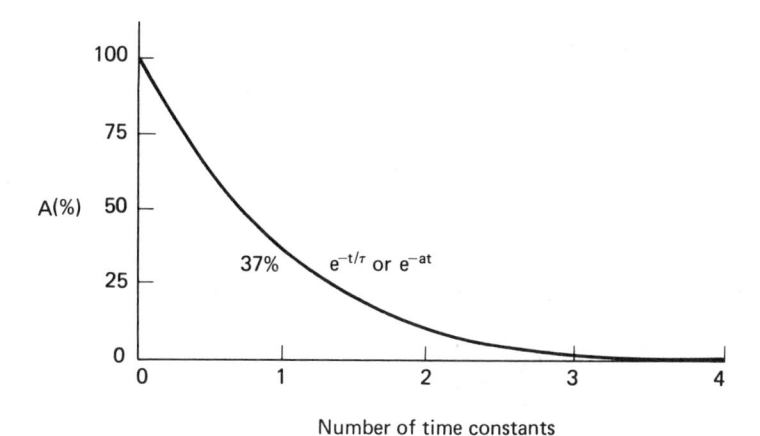

FIGURE F-1 Decaying exponential function.

replaced by $(s + a)$. Then

$$F(s) = \frac{1}{s + a} \text{ for } f(t) = e^{-at} \quad \text{(F-4)}$$

A popular form of equation (F-4) is readily obtainable if a is replaced by $1/\tau$ making $f(t) = e^{-t/\tau}$

$$\text{and } F(s) = \frac{1}{s + 1/\tau} = \tau \cdot \frac{1}{(s\tau + 1)} \quad \text{(F-5)}$$

Several of the more readily recognizable responses as a function of time are shown and the $f(t)$ to $F(s)$ are presented in Figure F-2. These and other commonly used transformations from $f(t)$ to $F(s)$ or from $F(s)$ to $f(t)$ are given in Table 4-1.

TABLE F-1 Laplace mathematical operations.

Operation	$f(t)$	$F(s)$
Addition	$f_1(t) + f_2(t)$	$F_1(s) + F_2(s)$
Multiplication	$f_1(t) \cdot f_2(t)$	$F_1(s) \cdot F_2(s)$
Multiplying by a constant	$K \cdot f(t)$	$K \cdot F(s)$
Differentiation	$\frac{d}{dt} f(t)$	$s \cdot F(s)$
Integration	$f(t) \, dt$	$\frac{1}{s} \cdot F(s)$ or $\frac{F(s)}{s}$

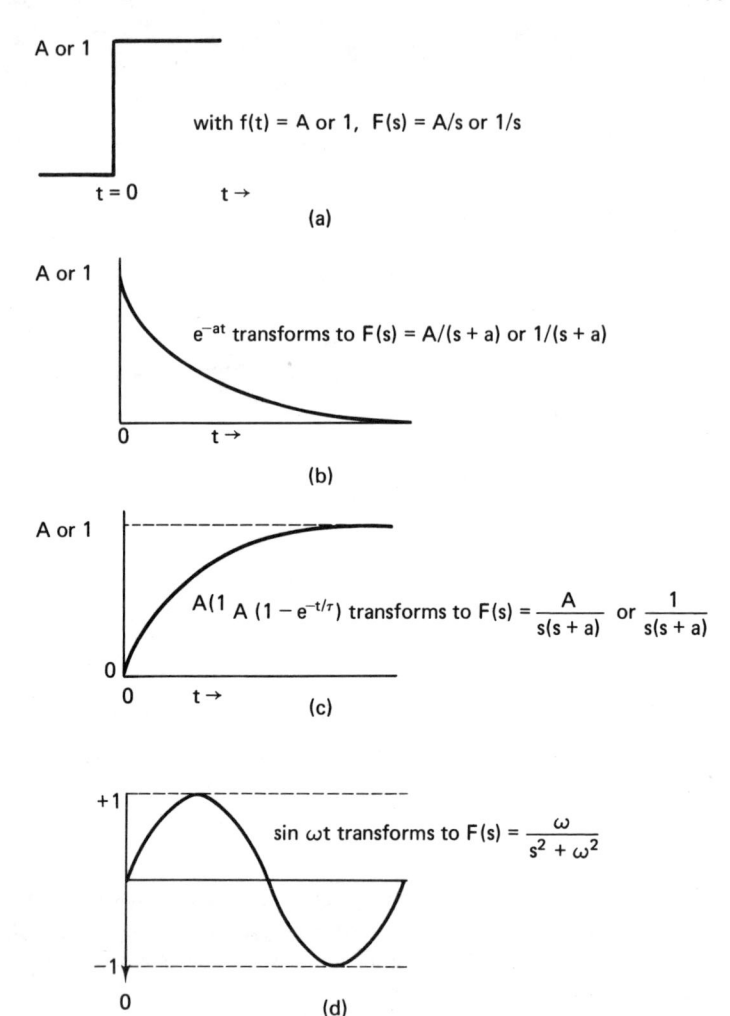

FIGURE F-2 Some basic transforms from $f(t)$ to $F(s)$: (a) a step input of amplitude A or 1; (b) the decaying exponential function; (c) the rising exponential function; (d) the sine wave.

(a) with $f(t) = A$ or 1, $F(s) = A/s$ or $1/s$

(b) e^{-at} transforms to $F(s) = A/(s+a)$ or $1/(s+a)$

(c) $A(1 - e^{-t/\tau})$ transforms to $F(s) = \dfrac{A}{s(s+a)}$ or $\dfrac{1}{s(s+a)}$

(d) $\sin \omega t$ transforms to $F(s) = \dfrac{\omega}{s^2 + \omega^2}$

A very important factor is the ease with which the Laplace transforms can be mathematically manipulated. Table F-1 shows the Laplace mathematical operations. Of specific interest is integration, which is simply the multiplication of a function of s by $1/s$ or differentiating which is simply multiplying a function of s by s.

Index

A

Accumulator, 105
Active devices, 82
Active network, defined, 91
Amplifiers, 8-11
   Bode plot for, 44-46
   differentiating, 87-88
   with feedback, *see* Feedback
   isolation, 17-18, 26
   operational, 2, 16-19, 86-90
   summing, 12, 18-20, 27
   transfer function for, 111
Amplitude (M_p), defined, 160
Analog computer, 86-90
   defined, 91, 215
Angular-displacement servo
   motor, 104
Automatic control systems, 1-2
   defined, 11

B

Bandwidth:
   of closed-loop system, 155-159, 174-176
   defined, 160
Beta, defined, 26
Block diagrams, 2
   of closed-loop system, 20-23
   of dc position control system, 126
   defined, 26
   of position control system, 209
Bode plot, 43-45
   of closed-loop system, 145-146
   defined, 4, 9
   of lead network, 82
   of open-loop system, 144
   of position control system:
      transfer function, 124-130

246 *Index*

Bode plot *(cont.):*
 of special transfer functions, 84-86, 90

C

Closed-loop-position control system, 142-143
Closed-loop system, 6, 8
 analysis of, *see* Closed-loop system analysis
 block diagram of, 20-23
 defined, 11
 transfer functions of, *see* Transfer functions
Closed-loop system analysis, 141-160
 closed-loop position control system, 142-143
 damping factor, 149-159
 effect of gain on closed-loop response, 147-149
 glossary, 159-160
 Nichols chart, *see* Nichols chart
 steady-state analysis, 144-146
 transient analysis, *see* Transient analysis
 See also Graphical solution of network response
Command signal, 106
 defined, 11
Comparator, defined, 11
Control system components, 95-117
 demodulator, *see* Demodulator
 drive motors, 104-107
 glossary, 116-117
 modulator, *see* Modulator
 phase compensation, 113-116
 position potentiometer, *see* Position potentiometer
 position transformer (synchro), 99-101
 rate generator, 114-116
 simple control system, flow diagram of, 112-113

Control system *(cont.):*
 transfer functions for, 95-96, 107-112
Control system concepts, basic, 1-12
 closed-loop system, 6, 8
 glossary, 11-12
 simple systems, 2-5
 time constant, 3-5
 transfer functions, 2, 8-11
Control winding, 104
Corner frequency, 33-36
 defined, 49
Critical damping, 183
 defined, 198

D

Damped frequency, 188
 defined, 199
Damped response to unit step input, 188-191
Damped wave, defined, 72
Damping, 181-183
 critical, 183, 198
 defined, 198
 viscous, 110, 206, 207, 208, 219-222
Damping coefficient, 71
 defined, 72, 198
Damping factor, 148-159, 182-184, 186-191
 defined, 160, 198
Dead space, 130-132
 defined, 137
Decade, 37
 defined, 49
Decibel, 32-33
Demodulator, 101-103
 defined, 116
 transfer function for, 111
Design considerations of dynamic systems, 203-215
 basic servo system, 209-215
 glossary, 215
 motor transfer function, 204-208

Differentiation, 87-88, 116
Drive motors, control system, 104-107
Dynamic systems:
 defined, 199
 design considerations of, see Design considerations of dynamic systems

E

Electrical resonant system, 194-195
Error, 2, 6, 8
Error signal, defined, 11
Exponential function, 5
 defined, 11

F

Feedback, 6, 15-27
 block diagram of, 20-23
 defined, 11
 glossary, 26-27
 isolation amplifiers and, 17-18
 negative, 17, 20-22, 24, 27, 31
 operational amplifiers and, 16-19
 phase shift, 25-26, 38, 40, 47-48
 position-control servo system with, 209-215
 positive, 21, 23-25, 27
 rate, see Rate feedback
 summing junction and, 18-20, 27
 See also Closed-loop system; Open-loop system
Feedback ratio (H), 142-143
 defined, 159
First-order system, 77
Forward gain (K), 122-130, 142-143, 171-174
 defined, 160
Frequency characteristics, 122, 142-143
 defined, 159

Frequency function, defined, 72
Function of time, defined, 72-73

G

Gain:
 defined, 177
 effect on closed-loop response, 147-149
Gain crossover frequency, 47-48, 82
 defined, 49
Gain effects on system performance, 130-133
Gear train, transfer function for, 111
Graphical solution of closed-loop systems, *see* Nichols chart
Graphical solution of network response, 31-49
 Bode plot, *see* Bode plot
 decade, 37, 49
 decibel, 32-33
 gain crossover frequency, 47-48, 49
 glossary, 49
 octave, 36-37, 49
 universal plot, 38-42

H

"Hunting," 137

I

Impedances, in operational form, 59, 60
Inertia:
 moment of, 206-208, 215, 219-222
 of the motor, 110
Input command, 2, 6
Instability, 24
Integrating amplifier, defined, 91

248 Index

Isolation amplifier, 17-18
   defined, 26

J

Jitter, defined, 137

L

Lag networks, 54
   defined, 27
   universal plot of, 38-42
Laplace transforms, 2, 58, 62-72, 241-244
   basic use of, 66-70
   complex transfer functions, solutions of, 77-90
   defined, 73
   series *LCR* circuit solution with, 70-72
   theory of, 63-65
   transient analysis using, 193-197
   See also s operator
Lead network, 54-56, 113-114, 229-231
   analysis of, 80-84
   defined, 91
   effect on phase margin, 133-135
Linear-displacement type servo control valve, 105-107
Linear synchro, 101

M

Modulator, 101-103
   defined, 116
   transfer function for, 111
Moment of inertia, 206-208, 219-222
   defined, 215
Motor time constant, 4-5
Motor torque constant, 108, 110
Motor transfer function:
   in terms of moment of inertia and viscous damping, 219-222

Motor transfer function *(cont.)*:
   in terms of speed/torque characteristics, 204-208
Motor velocity, transfer function of, 108

N

Natural frequency, 186-187
   defined, 199
Natural resonance, 150-151
   defined, 160
Negative feedback, 17, 20-22, 24, 31
   defined, 27
Networks:
   basic types of, 54-58
   solution of, *see* Graphical solution of network response; Laplace transforms; *s* operator
Nichols chart, 163-177
   defined, 177
   interpretation of, 169-177
   procedure for using, 164, 167-169
Normalizing, 40-42
   defined, 49

O

Octave, 36-37
   defined, 49
Open-loop system, 6, 7
   analysis of, *see* Open-loop system analysis
   defined, 11
   transfer functions of, *see* Transfer functions
Open-loop system analysis, 121-138
   defined, 138
   effect of lead network on phase margin, 133-135
   effect of rate feedback on phase margin, 135-136

Open-loop system analysis *(cont.):*
  gain effects on system
    performance, 130-133
  glossary, 137-138
  position control system
    transfer function, 123-130
  velocity lag error, 136-137
Operational amplifiers, 2, 16-19,
  86-90
Oscillation, 24
  period of, 186
Output response, 2
Overshoot, 186-187, 189-190
  defined, 199

P

Passive devices, 82
Peak frequency response M_p,
  147-149, 174-176
  defined, 160
Peak frequency response W_p,
  151-155
  defined, 160
Period of oscillation, 186
Phase compensators, 113-116
  defined, 116
Phase margin, 25, 38, 47, 233-235
  defined, 49, 138
  effect of lead networks on,
    133-135
  effect of rate feedback on,
    135-136
Phase shift, 25-26, 38, 40, 47-48
Position-control system, 209-215
  defined, 138
  equations, 143
  transfer function, 123-130
Position potentiometer, 97-99
  defined, 116
  transfer function for, 111
Position transformer (synchro),
  99-101
Positive feedback, 21, 23-25
  defined, 27
Proportionality constant, 136
  defined, 138

R

Rate feedback, 223-228
  defined, 138
  effect on phase margin, 135-136
Rate generator, 114-116
  defined, 116
Reactances, in operational form,
  59, 60
Rectilinear potentiometer, 98-99
Resonant circuit, 57-58
Resonant systems, types of,
  192-193
Response time, 184-185
  defined, 199
Rotary encoder, 101
Rotary position potentiometer, 98
Rotating system, 196-197

S

Second-order system, 77-80
  defined, 91
Series *LCR* circuit, 70-72
Series resonant circuit, 57-58
Servo control valve, 105-107
Servo motors, transfer function
  of, 107-110
Servo system, 8
  defined, 12
  transient analysis of, 197-198
Settling time, 186
  defined, 199
Simple motor systems, 2-5
Sine wave, 33
s operator, 2, 58-62
  basic use of, 59-62, 66-70
  complex transfer functions,
    solutions of, 77-90
  defined, 73
  series *LCR* circuit analysis,
    70-72
  See also Laplace transforms
Speed control system, 237-239
Spring-mass system, 195-196
Steady-state analysis, 144-146
Step input, 3-4, 70

Step input *(cont.):*
  defined, 73
  effect on system response, 180-191
Summing junction, 18-20
  defined, 12
Synchro, 99-101
Synchro control transformer, 100
Synchro generator, 99
Synchro motor, 100
Synchro pair, 99-100
Synchro receiver, defined, 116
Synchro transmitter, defined, 116

T

Tachometer, 114-116
  defined, 12
  transfer function for, 110, 111
Time constant, 3-5, 34-36
Time delay, 25-26
Torque, 206, 208, 220-222
  defined, 215
Torque constant of a motor, defined, 117
Transducer, defined, 27
Transfer characteristics, 9
Transfer functions, 2, 8-11
  of control system components, 95-96, 107-112
  defined, 12
  for position control system, 123-130

Transfer functions *(cont.):*
  solutions of, *see* Laplace transforms; *s* operator
Transient analysis, 179-199
  basic servo system, 197-198
  defined, 73
  effect of step input on system response, 180-191
  glossary, 198-199
  Laplace transforms and, 193-197
  types of resonant systems, 192-193
Transient response to a step input, 70
Type O system, 77, 142
Type 1 system, 77, 141-142
  defined, 91

U

Universal exponential curve, 5, 7
Universal plot, 38-42

V

Velocity constant of a motor, defined, 117
Velocity lag error, 136-137
  defined, 138
Viscous damping, 110, 206, 207, 208, 219-222
  defined, 215
Voltage amplifiers, 10, 15